Artificial Intelligence and IoT for Cyber Security Solutions in Smart Cities

This book offers a comprehensive overview of the current state of cybersecurity in smart cities and explores how AI and IoT technologies can be used to address cybersecurity challenges. It discusses the potential of AI for threat detection, risk assessment, and incident response, as well as the use of IoT sensors for real-time monitoring and data analysis in the context of smart cities. It includes case studies from around the world to provide practical insights into the use of AI and IoT technologies for enhancing cybersecurity in different contexts and highlight the potential benefits of these technologies for improving the resilience and security of smart cities.

Key Features:

- Studies the challenges of and offers relevant solutions to using AI and IoT technologies in cybersecurity in smart cities
- Examines the unique security risks faced by smart cities, including threats to critical infrastructure, data privacy and security, and the potential for large-scale cyber-attacks
- Offers practical solutions and case studies to be used to inform policy and practice in this rapidly evolving field
- Discusses the Fourth Industrial Revolution framework and how smart cities have been a significant part of this manufacturing paradigm
- Reviews aspects of Society 5.0 based on intelligent smart cities and sustainable issues for the cities of the future

Postgraduate students and researchers in the departments of Computer Science, working in the areas of IoT and Smart Cities will find this book useful.

Artificial Intelligence and IoT for Cyber Security Solutions in Smart Cities

Edited by
Smita Sharma, Manikandan Thirumalaisamy,
Balamurugan Balusamy, and Naveen Chilamkurti

CRC Press is an imprint of the
Taylor & Francis Group, an **informa** business
A CHAPMAN & HALL BOOK

Designed cover image: ShutterStock

First edition published 2025
by CRC Press
2385 NW Executive Center Drive, Suite 320, Boca Raton FL 33431

and by CRC Press
4 Park Square, Milton Park, Abingdon, Oxon, OX14 4RN

CRC Press is an imprint of Taylor & Francis Group, LLC

Library of Congress Cataloging-in-Publication Data
Names: Sharma, Smita, editor. | Thirumalaisamy, Manikandan, editor. |
Balusamy, Balamurugan, editor. | Chilamkurti, Naveen, 1974- editor.
Title: Artificial intelligence and IoT for cyber security solutions in smart cities / edited by Smita Sharma,
Manikandan Thirumalaisamy, Balamurugan Balusamy, Naveen Chilamkurt.
Description: First edition. | Boca Raton : C&H/CRC Press, 2025. |
Includes bibliographical references and index.
Identifiers: LCCN 2024030881 (print) | LCCN 2024030882 (ebook) |
ISBN 9781032605968 (hbk) | ISBN 9781032605999 (pbk) | ISBN 9781003459835 (ebk)
Subjects: LCSH: Smart cities—Security measures. |
Computer networks—Security measures. | Artificial intelligence—Industrial
applications. | Internet of things—Industrial applications.
Classification: LCC TD159.4 .A694 2025 (print) | LCC TD159.4 (ebook) |
DDC 307.1/416—dc23/eng/20240907
LC record available at https://lccn.loc.gov/2024030881
LC ebook record available at https://lccn.loc.gov/2024030882

ISBN: 9781032605968 (hbk)
ISBN: 9781032605999 (pbk)
ISBN: 9781003459835 (ebk)

DOI: 10.1201/9781003459835

Typeset in Times
by codeMantra

Contents

Editor Biographies

Dr. Smita Sharma is a senior IEEE member and a member of the Executive Committee IEEE UP Section. She has completed her PhD degree from Uttarakhand Technical University (Central Govt University) in the field of Wireless Body Area Sensor Networks. She received her BTech from Galgotias College of Engineering and MTech from Madan Mohan Malviya Engineering College, Uttar Pradesh, India, in Electronics and Communication Engineering. She is currently working at NIELIT, India. She has published over 35+ articles in peer-reviewed journals and conferences of international repute. Apart from that, 04 book chapters and 03 Indian patents are under her credit. She has edited various books and collaborated with eminent professors worldwide from top QS-ranked universities. Her current research interests include IoT, wireless sensor networks, network security, and AI. Her current focus in WSN is improving the lifetime and efficiency of sensor networks. She serves as a reviewer of many peer-reviewed journals and conferences. She is the organizer for several IEEE conferences like ICIPTM2022, ICIEM2022, ICCAKM2022, ICTACS 2022, IC3I 2022, IC3I 2023, UPCON2023, and IC3SE 2024. She has given several talks at various events and symposiums. She has also served on the publication team of several conferences of internationally reputed journals and is a member of IAENG and CSI societies. Her main research interest is in wireless sensor networks, the Internet of Things, artificial intelligence, and machine learning.

Prof. Balamurugan Balusamy is currently working as an associate dean student at Shiv Nadar University, Delhi-NCR. Prior to this assignment, he was a professor at the School of Computing Sciences and Engineering and Director of International Relations at Galgotias University, Greater Noida, India. His contributions focus on engineering education, blockchain, and data sciences. His academic degrees and 12 years of experience working as a faculty member at a global university like VIT University, Vellore, have made him more receptive and prominent in his domain. He does have 200 plus high-impact factor papers in Springer, Elsevier, and IEEE. He has written more than 80 edited and authored books and collaborated with eminent professors across the world from top QS-ranked universities. He has served up to the position of associate professor in his stint of 12 years of experience with VIT University, Vellore. He had completed his bachelor's, master's, and PhD degrees from premier institutions in India. His passion is teaching and adapting different design thinking principles while delivering his lectures. He has published 80+ books on various technologies and visited 15+ countries for his technical course. He has several top-notch conferences in his resume and has published over 200 quality journals, conferences, and book chapters combined. He serves in the advisory committee for several start-ups and forums and does consultancy work for the industry on industrial IoT. He has given over 195 talks at various events and symposiums.

Dr. Manikandan Thirumalaisamy is working as a professor in the Department of CSBS, Rajalakshmi Engineering College, Chennai, India. He holds a PhD degree in Information and Communication Engineering at Anna University Chennai, India. He secured a Master of Engineering in the CSE Department at Anna University, Chennai, Tamil Nadu, India. He graduated with a Bachelor of Engineering at Madurai Kamaraj University, Madurai, India. He has been in the teaching profession for more than 18 years. He has published 15+ research papers in various journals and conference proceedings. His research and publication interests include networks, cybersecurity, intrusion detection systems, computational intelligence, wireless networks, and congestion control. He is a senior member of the IEEE, ACM member, and CSI life member.

Professor Naveen Chilamkurti is the head of the Cybersecurity Discipline and associate dean of International Partnerships at La Trobe University, Melbourne, Australia. He also received the Australia-India Cybersecurity Infrastructure Grant, jointly funded by DFAT and the Indian government. He is a keynote speaker at various international conferences and has been recently elected as an IET (UK) fellow. He has an extensive research record in cybersecurity. He has published 360 journals/conference articles in cybersecurity, IoT, anomaly detection in IoT, the Internet of Medical Things, wireless security, federated learning in IoT, wireless multimedia, wireless sensor networks, and software-defined networks. He is active in editing and authoring nine books with reputed publishers. He has successfully attracted 20 research grants since 2000 to support PhD scholarships, fellowships, and travel grants for research collaboration. Prof. Chilamkurti secured 24 competitive grants from various sources, including SMART SAT CRC, Data61/CSIRO, Defence Science Institute, Australian Academy of Science, and OPTUS telecommunications. He was instrumental in designing and developing cybersecurity micro-credentials and other short courses now delivered online. He has taught networking, security, and cyber areas for 26 years and has supervised 51 research students to complete their master's and PhD programmes. He is also the director of La Trobe Cybersecurity Innovation Node, primarily focused on research, upskilling, and certification in cybersecurity programmes.

Contributors

A. Akshaya
Department of Computer Science and Engineering
Thiagarajar College of Engineering
Madurai, India

A. Anitha
Department of Software and System Engineering
Vellore Institute of Technology
Vellore, India

S. Anusuya
Saveetha Institute of Medical and Technical Sciences
Chennai, India

Saurav Bhattacharjee
Department of Engineering
National Institute of Technology
Silchar, India

K. Deepikakumari
V.S.B. Engineering College
Karur, India

K. Dharani
Department of Computer Science and Engineering
V.S.B. Engineering College
Karur, India

Purnima Gupta
Department of Computer Science Engineering
IMS-Ghaziabad University Courses Campus
Ghaziabad, India

T. Haritha
Vellore Institute of Technology
Vellore, India

M. Kalaiyarasi
Institute of AI & ML
Saveetha Institute of Medical and Technical Sciences, SIMATS
Chennai, India

Gitima Kalita
Kamrup Polytechnic,
Baihatachariali,
Assam, India

S. Karthi
Department of Computer Science and Engineering
V.S.B. Engineering College
Karur, India

V. Karthika
Department of Computer Science and Engineering
V.S.B. Engineering College
Karur, India

K. Kavya
Department of Computer Science and Engineering
V.S.B. Engineering College
Karur, India

Mukesh Kondala
Department of Operations and Applied Sciences
GITAM Deemed to be University
Visakhapatnam, India

Nisha Kumari
Department of Entrepreneur
GITAM Deemed to be University
Visakhapatnam, India

Raja Lavanya
Department of Computer Science and Engineering
Thiagarajar College of Engineering
Madurai, India

B. Malarvizhi
Saveetha Institute of Medical and Technical Sciences
Chennai, India

Atheer Abdullah Mohammed
Department of Industrial Management
Baghdad University
Baghdad, Iraq

N. R. Naveen Babu
Department of Computer Science and Engineering
Thiagarajar College of Engineering
Madurai, India

Logobe Riame
Kamrup Polytechnic
Baihatachariali, India

N. Sai Arunaa Varshini
Department of Computer Science and Engineering
Thiagarajar College of Engineering
Madurai, India

S. Saumya
Department of Computer Science and Engineering
V.S.B. Engineering College
Karur, India

S. Mercy Shalinie
Department of Computer Science and Engineering
Thiagarajar College of Engineering
Madurai, India

Archana Sharma
Department of Computer Science
Delhi Technical Campus
Greater Noida, India

Aswani Kumar Singh
Soft-tech Development Solution
Chandauli, India

Parampreet Singh
The Guardian Life Insurance Company of America
Mechanicsburg, Pennsylvania, United States

K. Sundarakantham
Department of Computer Science and Engineering
Thiagarajar College of Engineering
Madurai, India

M. Vasudevan
Department of Computer Science and Engineering
V.S.B. Engineering College
Karur, India

T. Vedhanayaki
Department of Computer Science and Engineering
V.S.B. Engineering College
Karur, India

1 Utilization of Artificial Intelligence and the Internet of Things in the Renewable Industry

Saurav Bhattacharjee, Logobe Riame, and Gitima Kalita

1.1 INTRODUCTION

Monitoring and data analysis in real time can improve how well energy is stored and traded. Energy storage is a key part of making sure that renewable power sources can meet both the supply and demand for power. It makes it possible to store extra energy and use it when demand is high. Industry 4.0 technologies can help make energy storage systems more efficient by using artificial intelligence (AI) and advanced data analytics to track and predict energy demand and supply. Optimization methods have evolved from math heuristics to metaheuristics and AI-based strategies; even though arithmetic heuristic techniques are capable of handling complicated problems of any size, they are sensitive to contemporary issues due to their inability to consider non-linearity limitations and issues like the emergence of high dimensionality. The meta-heuristic approach, sometimes known as a "nature-inspired algorithm," is superior to the math heuristic approach [1]. Decentralized energy trading platforms can also be made so that local groups can sell any extra power they make from renewable sources. There is an added advantage to storing renewable energy, which can lead to a more reliable and sustainable energy source. It is part of solving the problem of energy sustainability.

Additionally, with the support of 4.0 industry technological advances, industrial and manufacturing processes may grow more energy-efficient as a whole. In addition, improving the performance of electricity distribution, a smart microgrid enables the provision of energy management services through the utilization of big data analysis and intelligent sensors [2]. Optimization of energy usage and reduction of energy waste are made possible by the use of smart sensors and AI algorithms. As a result, this has the potential to aid international efforts to curb the release of greenhouse gases by the industrial sector. In addition to enhancing the dependability and stability of the energy system, the introduction of Industry 4.0 could help the inclusion of energy from renewable sources into the infrastructure. It can aid in the creation of smart charging stations that are equipped to run on renewable energy in response to

DOI: 10.1201/9781003459835-1

the rising popularity of electric cars. This has the potential to drastically cut down on transportation's carbon footprint, making for a greener tomorrow.

Zhou and Lund [3] studied a designed peer-to-peer process that can help prosumers, consumers, retailers, and aggregators that will benefit economically from sharing access to decentralized power sources. The carbon and graphite market are expected to grow from an anticipated $491 million in 2023 to $770 million by 2028, representing a CAGR of 9.4%. The study claims that factors like the increasing use of renewable energy in power production, falling prices for energy storage, and expanding availability of home storage systems are driving market growth. In this chapter, we explore at how energy management systems can be used in Industry 4.0 settings to reduce energy waste. The digital transformation of industry, often known as Industry 4.0, is critical to the progress of renewable energies.4.0 basically referred to the fourth revolution in industry and utilization of digital technology such as the Internet of Things, big data and system for cyber, and these technologies are implemented to improve the productivity and performance of various industries [4]. Scharl and Praktiknjo [5] analysed the implementation of 4.0, which can facilitate resource implementation through more versatility, enhanced accountability, improved effectiveness, and less use of energy. Improved maintenance practices can help renewable energy technologies operate better for longer. Predictive maintenance is an increasingly popular approach in the renewable energy sector that aims to lower operations and maintenance (O&M) costs. Furthermore, the renewable energy sector employs artificial intelligence (AI), big data analytics, machine learning, and the Internet of Things (IoT) to effectively analyse and monitor real-time data obtained from sensors in the practice of modelling equipment behaviour to facilitate the timely identification of damage, enables the monitoring of performance, allows for operational optimization, and aids in the prevention of failures and inefficiencies [6–10]. However, the incorporation of digital technology in the renewable energy industry presents several disadvantages stemming from the challenges associated with this industry which includes the potential risks to cyber-security, the scarcity of a proficient workforce, the requirement for significant financial investment, and the reluctance of workers to embrace change [11].

1.1.1 History of Industrial Revolution

There are six revolutionary stages of industry as described in Figure 1.1. As the term Industry 1.0 pertains to the era spanning the late 18th century and early 19th century,

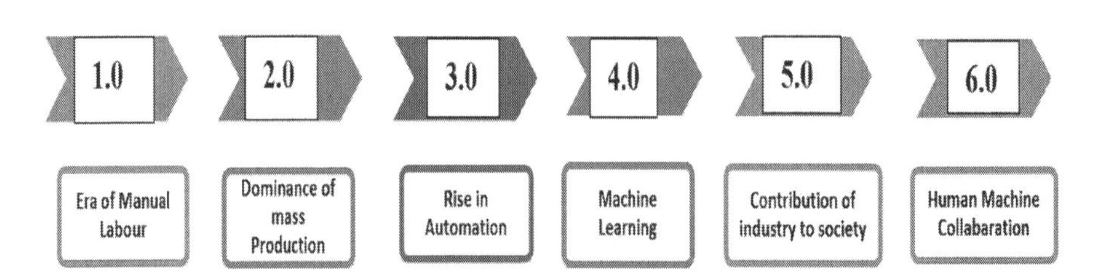

FIGURE 1.1 Stages of industrial revolution.

characterized by the predominant utilization of manual labour as the primary means of production. The era in question was distinguished by the emergence of the steam engine, which facilitated the automation of manufacturing processes. Industry 2.0 is the time period of the first half of the 20th century, when mass production methods were put into place. During this time, the assembly line was created, which made it easier to make a lot of goods at once and machines were used a lot more because electricity was becoming more common. Industry 3.0 denotes the era of the other remaining 20th century, characterized by the advent of automation and computerization. The era was distinguished by the extensive utilization of computers and the mechanization of numerous production procedures. The Hannover expo, the world's largest industrial trade expo, presented Industry 4.0 in 2011. Digital technology and the IoT are integrated into production in 4.0. Industry 4.0 uses AI, ML, and big data to enhance manufacturing processes. The subsequent stages of Industry 5.0 and 6.0 were briefly discussed thereafter. The advent of novel automation such as machine learning and artificial intelligence presents a promising avenue for tackling the challenges associated with the intermittency and ambiguity of renewable energy sources.

For several years, sustainability researchers and representatives of civil society have consistently emphasized the necessity of lifestyle modifications that align with the trajectory of digitization in which [12] investigated on the four primary impacts of digitization, which encompassed the direct effect, efficiency and rebound effects, economic growth, and sectoral change. The findings of Ref. [13] provide support for the concerns raised by critical perspectives that anticipate significant employment reductions resulting from the huge automation of manufacturing processes in Germany and China.

1.1.2 Algorithms for Renewable Energy Models

Alassery et al. [14] investigated three distinct solar prediction algorithms, namely Artificial Neural Networks (ANN), Support Vector Machines (SVM), and Random Forest (RF), for the purpose of forecasting solar energy and the findings indicate that the methodology achieved exceptional performance for the artificial neural network algorithm in comparison to the SVM and RF algorithms, with MAE values of 0.9558 and 1.7853 during the training and testing phases, respectively, and MFE values of 0.4456 and 0.5621 during the training and testing phases, respectively. The Model Predictive Control (MPC) algorithm is utilized to optimize the income maximization process [15]. Korjani et al. [16] recommend a strategy for managing and organizing the storage of energy from batteries in a microgrid that operates as a simulated power plant by utilizing time series prediction and artificial intelligence, specifically the neural network forecast regarding the combined load, PV generation, and shared services of a renewable energy community (REC) are utilized as inputs for optimization. By incorporating Industry 4.0 technology, production efficiency may be enhanced and a reliable power supply to the network can be maintained through the utilization of diverse energy resources such hydroelectric power, solar power, offshore wind power, geothermal power, and biomass fuel.

Furthermore, the implementation of IoT technology has the potential to address supply chain issues through its emphasis on optimizing the effectiveness, reliability, and security of power generation systems. The implementation of real-time monitoring has the potential to enhance operational efficiency by ensuring continuous monitoring of both demand and supply.

An intelligent grid based on (IoT) can facilitate remote and comprehensive monitoring of the grid. This particular smart grid has the capability to efficiently monitor variations in production, consumption, and supply. The utilization of cloud-hosted agents allows for remote monitoring of network operational activities in real time. The utilization of fog-based IoT and intelligent gateways has the potential to facilitate instantaneous monitoring of energy production stages by integrating wireless communication modules with sensors. The intelligent gateway obtains sensory information from individual plants via the wireless connectivity module and has the capability to predict electricity demand, generation, and cost. The rapid development of automated systems has encouraged power companies and consumers equally to embrace fog computing as a valuable asset for data processing [17]. The market in North America is anticipated to dominate the simulated power plant marketplace, with the United States as its key market. Currently, the Northern American region dominates a substantial portion of about 37.6% of the global market for virtual power plants. Collectively, both of these areas account for over 66% of the market [18]. Virtual assistants help customers in performing a diverse range of tasks, including but not limited to home automation and generating reminders. Commonly, they serve as a constituent element within a healthcare framework. The integration of a smart device with a virtual assistant facilitates the effortless transformation of informal language into a specialized vocabulary that is appropriate for both healthcare practitioners and individuals seeking medical information. It has the potential to facilitate the coordination of medical procedures, resulting in time-saving benefits for patients. Virtual assistants have the potential to increase productivity and decrease the required staff and resources in medical institutions. Figure 1.2 illustrates the

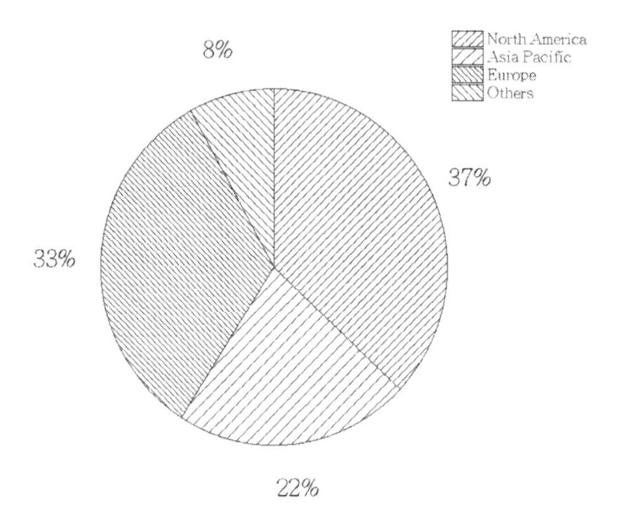

FIGURE 1.2 Pie chart illustrating the worldwide virtual power plant market [18].

worldwide virtual power plant market, highlighting the primary regions that contribute to its growth, namely North America, Europe, and Asia Pacific.

Magazzino et al. [19] employed a sophisticated estimation methodology within the field of advanced machine learning, integrating a direct-to-consumer causality model to assess the interrelationships between coal, wind, economic growth, and CO_2 emissions in which the target variable chosen for analysis is carbon dioxide (CO_2) emissions. Long-term estimations [20] have revealed that a marginal rise of 1% in the utilization of renewable energy sources is significantly linked to a concomitant reduction of 0.23% in the growth rate of real gross domestic product (GDP). In the study, Apergis and Payne [21] investigated the correlation among carbon dioxide, energy from nuclear consumption, energy from renewable sources consumption, and economic growth. The dataset utilized for analysis comprised 14 countries categorized as developed as well as 5 countries categorized as developing. Manage information using algorithms that can transfer data in a way that saves energy, combine communication and computation, and handle requests [22–24]. Sherazi et al. [25] studied about the utilization of energy harvesting techniques and the potential benefits it can offer in terms of enhancing the production efficiency of different product lines within the context of Industry 4.0 and improvement is contingent upon the specific values of cost of production and rate of production, which represent key parameters of the manufacturing plant.

1.2 ADVANCEMENT OF INDUSTRY 4.0 WITH RENEWABLE ENERGY

The utilization of 4.0 has the ability to improve the dependability of renewable energy generation. The integration of the Internet of Things, machine learning, and big data has the potential to optimize the renewable energy generation process and facilitate the establishment of a more sustainable circular economy within the framework of this industry.

Industry 4.0 technologies may eventually incorporate renewable energy sources through the implementation of virtual power plants and microgrids. The network is integrated with virtual power plant, which coordinates multiple decentralized power generators into a single unified system. The utilization of the network has the tendency to generate a reliable source of electrical power at the highest level of efficiency. In another aspect, microgrid are independent power system that operate separately from the main power grid but on a smaller scale. AI techniques are used to detect any anomalies in the system which can be grouped into two approaches machine learning and statistic-based approach [26]. The utilization of organic materials holds promise for enhancing current energy storage technologies, while simultaneously offering environmentally sustainable, adaptable, and potentially cost-effective energy storage devices [27]. The incorporation of AI in the energy industry has led to substantial growth as each company actively seeks to implement and manage energy demand equilibrium. They can also forecast the demand for energy needs through smart meters. Further, the algorithm also offers energy usage pattern to consumer for reducing expenses [28]. Internet of Things (IoT) system harnesses electricity from three distinct modules: a piezoelectric sensor, a body heat converter, and solar panels as these modules are interconnected to a power storage circuit, which facilitates the generation of electrical energy to establish the accuracy and reliability of the data produced by these modules [29], artificial neural network and adaptive neuro fuzzy interface system will be employed for the prediction of power output.

The renewable energy platform integrates several renewable energy sources, such as wind energy, using an IoT platform. The Internet of Things platform connects renewable energy sources to a virtual power plant, ensuring energy storage. The simulated power plant utilizes predicting machine learning and analytics algorithm to improve energy efficiency and waste minimization. Battery backup technological advances are utilized to store excess energy for subsequent utilization during periods of high demand. Ultimately, the dissemination of energy to end-users is facilitated by a sophisticated smart grid infrastructure that leverages cutting-edge sensors and monitoring mechanisms to optimize the dependable and efficient distribution of power.

Some investigators [30] performed a study using qualitative comparative fuzzy-set analysis (fsQCA) to investigate the consequences of digitization which will affect company performance frameworks and the overall satisfaction of various individuals. To classify energy consumers at manufacturing companies, an energy portfolio as well as focused on objectives and action plans to increase the production of energy has been presented [31]. Researchers [23] investigated the performance indicators and a digital data collection system to monitor the status of manufacturing procedures and the use of energy in real time. This approach enables the detection of typical energy usage throughout the operational cycle of a machine.

Figure 1.3 illustrates the use of sensor for the goal of acquiring information on renewable sources including turbines for wind and solar panels. The collected data is sent to an inverter, which converts the electrical energy from DC (direct current) to alternating current, or AC, for use in a microgrid. The battery storage system serves

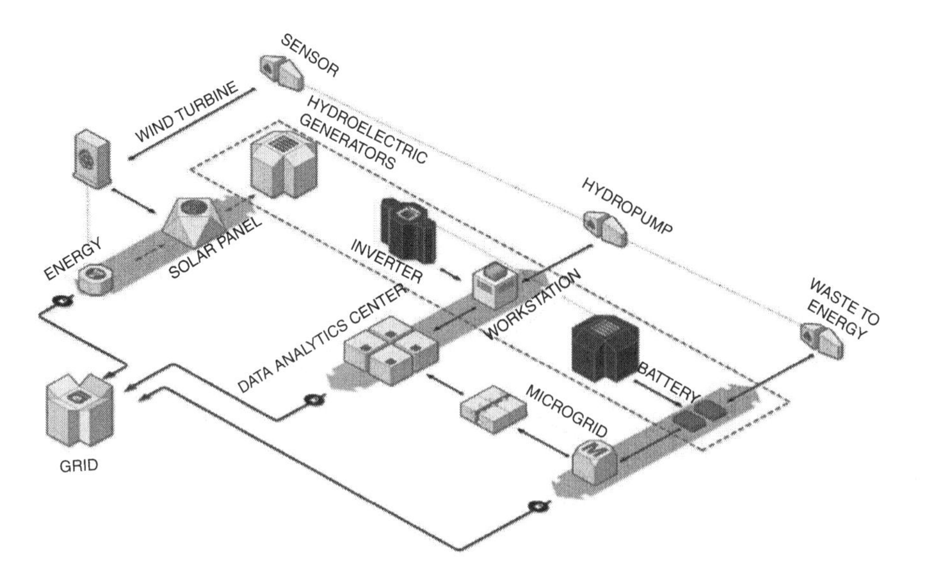

FIGURE 1.3 Renewable energy system architecture.

the purpose of storing surplus energy to be utilized during phases of reduced energy generation. The energy grid serves as a means of linking to the primary power grid for the purposes of energy exchange and dispersal. The system comprises a range of sustainable energy sources, including photovoltaic cells, wind turbines, and hydroelectric generators, all of which are linked to a digital power station. The virtual power plant employs (IoT) devices and machine learning algorithms to oversee and regulate the power distribution process, guaranteeing a consistent and dependable energy output. The system incorporates energy storage alternatives, such as batteries and pumped hydro storage, to accumulate surplus energy and furnish electricity during periods of insufficient supply [32]. The system ultimately incorporates a circular economy methodology through the utilization of waste-to-energy facilities and resource recycling to mitigate waste and enhance efficacy. The integration of renewable energy sources, utilization of IoT devices and machine learning algorithms, and implementation of a circular economy approach can enhance the reliability, efficiency, and sustainability of the renewable energy sector, thereby making it a viable source of power for the future. Incorporating Industry 4.0 technologies with renewable energy sources has the potential to enhance energy production and distribution in terms of efficiency, reliability, and sustainability.

1.2.1 Integration of Industry 4.0 with Renewable Energy System

The integration of Industry 4.0 technology has the potential to improve the efficiency, reliability, and longevity of renewable energy systems. Edge devices such as sensors, smart meters, and actuators are utilized for the purpose of data collection and control, while the cloud platform is employed for data processing and storage. The various components of the system, such as energy storage systems, AI/ML predictive analytics, and alternative energy sources, are interconnected through networks of communication. The energy market facilitates the effective transfer of energy between producers and consumers through the utilization of trading platforms and smart contracts.

Industry 4.0 endeavours to establish a renewable energy system that is decentralized, environmentally sustainable, and possesses the ability to adapt to dynamic circumstances in real time. The emergence of the Internet of Things (IoT), artificial intelligence (AI), and blockchain represents technological progress that has facilitated the transition towards decentralization and sustainability. The implementation of these technologies has enabled the energy sector to enhance its accountability, transparency, and efficiency. The fourth industrial revolution has the potential to change the approaches energy is generated, distributed, and consumed.

It is of the utmost importance to do a full cost-benefit analysis of the Industry 4.0 approach in terms of sustainability, considering the projection of the future. Undoubtedly, over the course of an extended period, it becomes evident that the adoption and implementation of this philosophy necessitates an initial cost. Currently, the philosophy is in its early stages and necessitates further elucidation and specialized development. Despite the initial expenditures, the long-term risk benefit analysis indicates that the fourth industrial revolution has enormous potential to produce an advantage for firms, assuring the production of long-term value [33]. Enterprises are positive about the benefits of the 4.0 revolution on three pillars namely economy, society, and environment. From an economic standpoint, Industry 4.0 is anticipated to alleviate the strain on utilized raw materials, diminish lead times, and establish sustainable production systems that prioritize energy efficiency. Industry 4.0 is anticipated to generate a workplace environment that is both secure and conducive to well-being, foster a comprehensive learning culture, and enhance the skill sets of employees. Ecologically speaking, the primary concerns pertain to the utilization of natural resources, the production of waste, and the effective management of non-renewable resources, with a particular emphasis on energy.

The advent of Industry 4.0 presents significant prospects for the creation of sustainable products that offer enhanced service functionalities to customers [34]. Strategic competitive edge is expected to provide various benefits to firms, including savings on supplies and energy preservation, adherence to rules and regulations, and improved procedures that result in greater quality and improved client satisfaction. However, it is important to acknowledge that these advantages are accompanied by potential drawbacks from an alternative viewpoint. Development as well as implementation of the fourth industrial revolution, has multiple issues that require attention and resolution [35]. Figure 1.4 illustrates a flow chart of a renewable system that showcases the application of predictive analysis through machine learning.

1.3 CASE STUDY

1.3.1 ReNew Power

Indian renewable energy business ReNew Power uses Industry 4.0 to optimize its wind and solar power installations. The company's objective is to sustain its growth by broadening its range of investments, enhancing the variety of renewable energy sources it utilizes, and venturing into untapped international markets. The commitment of ReNew Power to sustainable development and its capacity to adjust to market dynamics are expected to contribute to its sustained success within the renewable energy industry. IoT sensors, data analytics, and machine learning algorithms monitor and control renewable energy production and distribution, predict maintenance

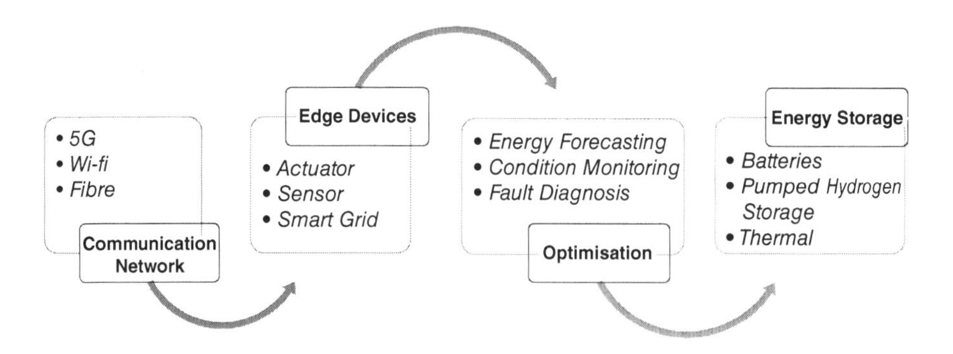

FIGURE 1.4 Application of renewable system depicting predictive analysis by machine learning.

needs, and reduce downtime. This improved renewable energy system efficiency and reliability. ReNew Power adheres to an asset-intensive operational framework, encompassing the development, ownership, and operation of large-scale renewable energy projects in the utility sector. The company places emphasis on the utilization of wind and solar energy, maintaining a diversified portfolio that spans various states within India. ReNew Power engages in various stages of the project lifecycle, encompassing activities such as site identification and acquisition, project development, construction, and operations. ReNew Power serves a major part in mitigating climate change and minimizing the release of greenhouse gas emissions through the production of clean and sustainable energy. The company operation has significantly reduced CO_2 emission by millions of tons each year. Renew power commitment to sustainability extends beyond power installation including programme such as reforestation and waste management. Renew power is in the position to capitalize India's growing need for renewable energy

1.3.2 Study on Suzlon Energy

Suzlon Energy, an Indian wind turbine manufacturing company, boosted the effectiveness and performance through machine learning in order to improve the design and performance of wind forecasting this company advances in machine algorithm and data analytics which increase the effectiveness. In order to gather information on wind speed, temperature, power output they have installed sensor on wind turbine. Suzlon energy has been able to forecast requirement, and it can find probable effect by advancing machine algorithm. Through the use of centralized control system industry is able to arrange depict the output of wind turbine from a single location. Suzlon energy has adopted augmented reality technology to boost the effectiveness of maintenance procedure for wind turbine. This business has created an augmented reality software that gives maintenance to remote access to staff which results in few on-site visits. Suzlon energy noteworthy expansion is of 25.00% compound annual growth rate for 2003–2008. In addition, Suzlon Energy has made investments in Industry 4.0 technologies with the aim of enhancing the efficacy and functionality of its wind turbines. This strategic move has the potential to lower expenses and enhance the company's competitiveness within the market.

1.4 APPLICATION OF FOURTH INDUSTRIAL REVOLUTION ON HEALTHCARE MANAGEMENT SYSTEM

In brief, the utilization of machine learning and big data analysis in healthcare can prove to be highly advantageous as it enables healthcare practitioners to scrutinize vast quantities of patient data, thereby facilitating the identification of patterns and predictions pertaining to patient outcomes. The utilization of this data has the potential to facilitate the delivery of tailored healthcare and interventions to individuals, ultimately resulting in improved health results. Then, the collected data is analysed to look for trends and correlations that could help predict how patients will do. After analysing of data, ML models are used for building forecast models. Using these models can help doctors make predictions about how a patient will do, such as how likely it is that they will get a certain disease or how well a certain treatment will work. In the realm of medical diagnostics and scientific exploration 4D scanning adds the dimension of time to imaging and offers a dynamic and comprehensive view of objects and structures. Furthermore, 4D MRI imaging allows for the assessment of organ function and blood flow offering critical information for neurological procedures. 4D scanning plays pivotal role in scientific research and exploration. In the field of biology researcher utilize 4D microscopy to observe cellular dynamics. In material science and engineering, 4D imaging techniques offer insights into the behaviour of complex material and structure under varying conditions.

The concept of smart healthcare encompasses various entities such as clinical and scientific research institutes, regional centres for health decision-making, individual or family users, and healthcare information systems [36–40]. Numerous scholarly research publications havc dcmonstrated the extensive array of artificial intelligence (AI) implementations within the healthcare industry. A number of healthcare applications have been created to aid in determining the existence of various ailments such as brain tumours and cancer [41–43]. Figure 1.5 illustrates how patient records can be kept in a database, which is the most important tool for big data analysis.

These applications utilize advanced imaging techniques such as mammography and magnetic resonance imaging to identify and predict the presence of these diseases. Intelligent healthcare establishments encompass medical facilities, physicians, and research institutions. Real-time monitoring and intervention can be employed to offer prompt feedback and treatment to patients, relying on the prognostications generated by the predictive models. Real-time monitoring of a patient's vital signs could enable the prompt initiation of preventive measures in the event of a high-risk prediction of a heart attack by a machine learning model.

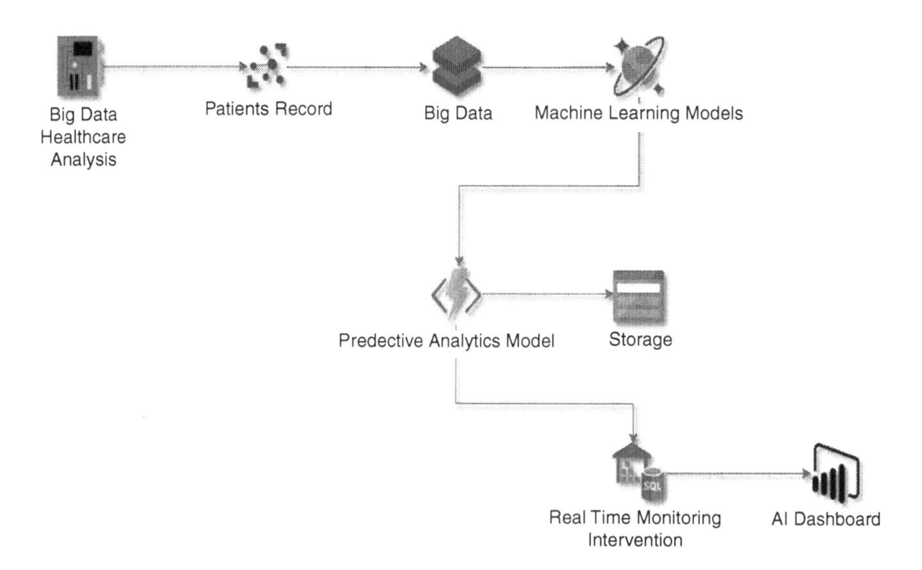

FIGURE 1.5 Machine learning health care architecture.

1.5 FUTURE TREND IN INDUSTRY AND SOCIETY 5.0

Industry 4.0 and Industry 5.0 refer to different stages of the revolution in industry. Society 1.0 is characterized as a collective of individuals who engage in hunting activities while maintaining a harmonious relationship with the natural environment. Society 2.0 emerged as a result of the agricultural revolution, which led to the development of complex social structures, increased organizational capabilities, and the establishment of nation-states while Society 3.0 is distinguished by a perspective that prioritizes industrialization as a means to achieve widespread production and Society 4.0 can be defined as a societal framework that primarily relies on information, enabling the creation of significant value through the interconnection of intangible resources, such as information networks so the 5.0 society can be characterized as a data-driven society that has evolved from the 4.0 society, which prioritized the well-being of individuals [44]. The advent of Industry 5.0 is anticipated to result in a notable change in outlook, wherein considerable importance will be accorded to individuals as the focal point of the manufacturing sector. In the forthcoming digital revolution, the commodities and amenities are expected to be customized in accordance with the specific demands of the consumers.

On the other hand, Industry 5.0 is a new idea that has not fully grown yet. Industry 5.0 suggested upgrade to Industry 4.0. It would combine the benefits of Industry 4.0 with the people-focused approach of Industry 3.0. The economic growth of a nation is contingent upon the widespread dissemination and assimilation of novel technologies, and in the process, Japan has established the concept of Society 5.0, which pertains to a contemporary society that utilizes connected objects, Big Data, and artificial intelligence to enhance the world [45,46]. The Industry 5.0 paradigm gives top priority to making a manufacturing process that puts more emphasis on human participation and encourages workers and machines to work together. Industry 4.0 is mostly about how technology can be used to automate and improve the manufacturing process. Industry 5.0, on the other hand, takes a more all-encompassing method that combines human intelligence and creativity with advanced technology. The goal of Industry 5.0 is to create a production system that is efficient, good for the environment, accountable from a social point of view, and helps people. Society 5.0 is anticipated to facilitate an interconnected environment in cyberspace, wherein various components such as individuals, objects, and systems are seamlessly integrated [47]. To sum up, Industry 4.0 and Industry 5.0 are two different parts of the industrial change. Industry 4.0 puts an emphasis on automation and technological progress, while Industry 5.0 puts an emphasis on a holistic approach that combines the benefits of technology with human contact. The development of both society and industry with the arrival of version 5.0 brings in a paradigm shift in the way we perceive technology, economics, and the dynamics of society. As we are on the verge of entering this transformative era, which is characterized by the combination of human-centred values with digital innovation, it is absolutely necessary for us to traverse this environment with vision and responsibility. It is possible for us to harness the full potential of novel technologies by embracing collaboration, inclusion, and ethical considerations. This will allow us to address worldwide issues, drive sustainable growth, and improve the quality of life for every segment of society. Most

TABLE 1.1
Recent Advancement in Industry 5.0

Development	Application	References
1. Sustainable supplier evaluation framework designed for the era of Industry 5.0	The research combines the methods of VC-DRSA, CRITIC, and a modified CTOPSIS	[48]
2. Optimizing the use of energy, materials processing, and product lifecycles in an environmentally friendly way, considering workers and society	Lifecycles in the Scopus database back up Industry 5.0's bibliometric research	[49]
3. Industry 5.0 has raised the importance of human-centred manufacturing. Manufacturing processes should prioritize worker well-being	The 5C evolution map shifts human-machine relationships from survival to partnership to empathy and shared evolution	[50]
4. Human creativity with intelligent efficient machine interface	New network speeds of 6G and beyond, as well as edge computing, digital twins, collaborative robots, the Internet of Things, and blockchain technology	[51]
5. Clean, affordable, and sustainable energy sources are needed due to fossil fuel shortages, which also includes production of algae	The utilization of genetic engineering techniques in the cultivation of algae for bioenergy production	[52]
6. Industry 5.0 continues to evolve to address the demands of its customers	Big data, smart sensors, multi-agent systems, complex adaptive systems, smart materials, 5D printing, 4D scanning	[53]
7. When humans and machines finally find common ground and work together in harmony, a new revolutionary wave known as Industry 5.0 will have crested in the "age of augmentation"	Principled framework called Value Sensitive Design (VSD) is proposed	[54]
8. Robots collaborate with the human brain and function as co-workers rather than rivals	Intelligent devices, systems, and automation seamlessly integrate with the physical world with human intelligence	[55]
9. In this emerging field, big data, analytics, and extensive storage are crucial, and synthetic biology has the potential to be as ubiquitous and transformative as digitization and the Internet	"White biotechnology" is used by chemical industries to develop new processes, find alternative raw materials, and conserve resources	[56]

of the developments in Industry 5.0 which can be seen in Table 1.1, are related to sustainable development and optimizing the use of energy and material processing.

1.6 CONCLUSION

In the final analysis, the energy integration from renewable energy sources infrastructure via artificial intelligence and machine learning requires an efficient energy management system (EMS) in order to optimize energy consumption. The EMS monitors and controls energy characteristics, predicts energy demand, and adjusts energy supply to decrease waste and maximize efficiency carbon footprint, lowering greenhouse gas emissions. Industry 6.0 is developing quickly, and it has the potential to aid in diagnosis and treatment, with an emphasis on health sector. Researchers face challenges in understanding the future orientation of Industry 5.0. There are multiple methodologies as discussed in this chapter that can be employed for the purpose of technology forecasting, future impacts analysis, or resilience efforts. The involvement of additional developing countries is of paramount importance in order to ensure the significant impact of 5.0 as an initiative focused on technological advancements in the renewable sector.

REFERENCES

1. M. Nuttaki and S. Mandava, "Review on optimisation technique and role of artificial intelligence ini home energy management system," *Engineering Application of Artificial Intelligence*, vol. 105721, p. 119, 2023.
2. D. A. Widodo, N. Iksan, E. D. Udayanti and Djuniadi, "Renewable energy power generation forecasting using deep learning method," *IOP Conference Series: Earth and Environmental Science*, Semarang, Indonesia, 2021.
3. Y. Zhou and P. D. Lund, "Peer-to-peer energy sharing and trading of renewable energy in smart communities: Trading pricing models, decision-making and agent-based collaboration," *Renewable Energy*, vol. 325, 2023.
4. D. Hernandez, G. Peralta, L. Manero, R. Gomez, J. Bilbao and C. Zubia, "Energy and coverage study of LPWAN schemes for Industry 4.0," In *Proceedings of the 2017 IEEE International Workshop of Electronics, Control, Measurement, Signals and Their application to Mechatronics (ECMSM)*, Donostia, Spain, 24–26 May 2017.
5. S. Scharl and A. Praktiknjo, "The role of a digital industry 4.0 in a renewable energy system," *International Journal of Energy Research*, vol. 43, pp. 3891–3904, 2019.
6. IEA, *Digitalization & Energy*, IEA, Paris, France, 2017.
7. W. Shin, J. Han and W. Rhee, "AI-assistance for predictive maintenance of renewable energy systems," *Energy*, vol. 221, 2021.
8. L. Bosman, W. Leon-Salas, W. Hutzel and E. Soto, "PV System predictive maintenance: Challenges, current approaches, and oppurtunities," *Energy*, vol. 13, 2020.
9. M. Canizo, E. Onieva, A. Conde, S. Charramendieta and S. Trujillo, "Real-time predictive maintenance for wind turbines using Big Data Frameworks," In *Proceedings of the 2017 IEEE International Conference on Prognostics and Health Management*, Dallas, USA, 2017.
10. J.-Y. Hsu, Y.-F. Wang, K.-C. Lin, M.-Y. Chen and J.-Y. Hsu, "Wind turbine fault diagnosis and predictive maintenance through statistical process control and machine learning," *IEEE Access*, pp. 23427–23439, 2020.
11. M. Sony, "Pros and cons of implementing Industry 4.0 for the organizations: A review and synthesis of evidence," *Production & Manufacturing Research*, vol. 8, pp. 244–272, 2020.
12. L. Steffen, J. Pohl and T. Santarius, "Digitalization and energy consumption. Does ICT reduce energy demand?" *Ecological Economics*, vol. 176, 2020.
13. G. Beiger, S. Neihoff, T. Ziems and B. Xue, "Sustainability aspects of a digitalized industry: A comparative study from China and Germany," *International Journal of Precision Engineering and Manufacturing-Green Technology*, vol. 4, no. 2, pp. 227–234, 2017.
14. F. Alassery, A. Ahmed, A. I. Khan, K. Irshad and S. Islam, "An artificial intelligence based solar radiation prophesy model for green energy utilization in energy management system," *Sustainable Energy Technologies and Assesment*, vol. 52, 2022.

15. C. Francesco, D. A. Federico, G. Natrella and M. Merone, "A new hybrid AI optimal management method for renewable energy communities," *Energy and AI*, vol. 10, 2022.
16. S. Korjani, A. Facchini, M. Mureddu, A. Rubino and A. Damiano, "Battery management for energy communities-Economic evaluation of an artificial intelligence led system," *Journal of Cleaner Production*, vol. 314, 2021.
17. B. Farahani, F. Firouzi, V. Chang, M. Badargolu and K. Makodiya, "Towards fog driven IOT Health: Promises and challenges of IOT," *Future Generation Computing System*, vol. 78, no. 2, pp. 659–76, 2018.
18. N. Sonnichsen, Virtual Power Plants Market Share Globally by Region, Statista.
19. C. Magazzino, M. Mele and N. Schneider, "A machine learning approach on the relationship among solar and wind energy production, coal consumption, GDP, and CO2 emissions," *Renewable Energy*, 2020.
20. C. Magazzino, "On the relationship between disaggregated energy production and GDP in Italy," *Energy and Environment*, vol. 12, no. 8, pp. 1191–1207, 2012.
21. N. Apergis and J. Payne, "Renewable energy consumption and economic growth from a panel of OECD countries," *Energy Policy*, vol. 38, pp. 656–660, 2010.
22. T. Javied, J. Bakakeu, D. Gessinger and J. Franke, "Strategic energy management in industry 4.0 environment," In *Annual IEEE International System Conference*, Vancouver, BC, Canada, 2018.
23. M. Jin, R. Tang, Y. Ji., F. Liu, L. Gao and D. Huisingh, "Impact of advanced manufacturing on sustainability overview of the special volume on advanced manufacturing for sustainability and low fossil carbon emission," *Journal of Cleaner Production*, 2017.
24. S. Wan, Y. Zhao, T. Wang, Z. Gu, Q. H. Abbasi and K. R. Choo, "Multi-dimensional data indexing and range query processing via voronoi diagram for internet of things," *Future Generation Computer System*, vol. 91, vol. 91, pp. 382–391, 2019.
25. H. H. Sherazi, L. A. Grieco, M. A. Imran and G. Boggia, "Energy-efficient LoRaWAN for industry 4.0 applications," *IEEE Transactions on Industrial Informatics*, vol. 17, no. 2, pp. 891–902, 2021.
26. J. Peppanen, X. Zhang, S. Grijalva and M. J. Reno, "Handling bad or missing smart meter data through advanced data imputation," *IEEE Power Energy Society*, Minneapolis, MN, USA, pp. 1–5, 2016.
27. P. Poizot, F. Dolhem and J. Gubicher, "Progress in all-organic rechargeable batteries using cationic and anionic configurations: Toward low-cost and greener storage solutions?" *Current Opinion in Electrochemistry*, vol. 9, pp. 70–80, 2018.
28. A. Kumari, R. Gupta, S. Tanwar and N. Kumar, "Blockchain and AI amalgamation for energy cloud management: Challenges, solutions, and future directions," *Journal of Parallel and Distributed Computing*, vol. 143, pp. 148–166, 2020.
29. V. Puri, S. Jha, R. Kumar, I. Priyadarshani, L. H. Son, A. Basset and H. V. Long, "A hybrid artificial intelligence and internet of things model for generation of renewable resource of energy," *IEEE Access*, vol. 7, 2019.
30. A. C. Llopis, F. Rubio and F. Valero, "Impact of digital transformation on the automotive industry," *Technological Forecasting and Social Change*, vol. 120343, p. 162, 2021.
31. S. D. Tesch, F. S. da Costa and C. C. Paredes, "Looking at energy through the lens of Industry 4.0: A systematic literature review of concerns and challenges," *Computers & Industrial Engineering*, vol. 143, 2020.
32. D. Zeevi, T. Korem and N. Zmora, "Personalised nutrition by prediction of glycemic responses," *Cell*, vol. 163, no. 5, pp. 1079–1094, 2015.
33. M. Muller, O. Julian, Buliga and K. I. Voigt, "Fortune favours the prepared," *Technological Forecasting and Social Change*, vol. 17, no. 2, p. 132, 2017.
34. M. Faheem, S. B. Shah, H. Butt, R. A. Raza, M. Anwar, M. W. Ashraf and V. C. Gungor, "Smart grid communication and information technologies in the perspective of Industry 4.0: Opportunities and challenges," *Computer Science Review*, vol. 30, pp. 1–30, 2018.

35. S. Avikal, A. Singh and M. Ram, *Sustainability in Industry 4.0: Challeneges and Remedies*, CRC Press, USA, vol. 1, 2021.
36. J. Redfern, "Smart health and innovation: Facilatating health related behaviour change," *Proceedings of the Nutrition Society*, vol. 76, pp. 328–332, 2017.
37. R. White, "Skill discovery in virtual assistant," *Communications of the ACM*, vol. 61, no. 11, pp. 106–113, 2018.
38. C. Ortiz, "Holistic conversational assistant," *AI Magazine*, vol. 39, no. 1, pp. 106–113, 2018.
39. P. Yang and W. Fu, "Mindbot: A social based medical virtual assistant," In *IEEE International Conference on Healthcare Informatics*, New York, 2016.
40. B. Liu and G. Zhi, "Impact of big data and artificial intelligence on the future medical model," *Technological Forecasting and Social Change*, vol. 165, no. 22 pp. 120557, 2018.
41. J. Martin, H. Varilly, J. Cohn and G. Wightwick, "Preface: Technologies for a smarter planet," *IBM Journal of Research and Development*, vol. 54, no. 4, pp. 1–2, 2010.
42. F. Gong, X. Sun, J. Lin and X. Gu, "Primary exploration in the establishment of chinas's medical treatment," *International Journal of Biological Macromolecules*, vol. 11, no. 2, pp. 28–29, 2013.
43. F. Pan, *Interviewee, Interview with Wu He-Quan*, [China Medical Herald Interview], 2019.
44. M. Tabaa, F. Monterio, H. Bensag and A. Dandache, "Green industrial Internet of Things from a smart industry perspective," *Energy Reports*, vol. 6, pp. 430–466, 2020.
45. T. Salimova, N. Guskova, I. Krakovskaya and E. Sirota, "From industry 4.0 to society 5.0: Challenges for sustainable competitiveness of Russian industry," *Material Science Engineering*, vol. 497, no. 1, pp. 012090, 2019.
46. B. S. Glaser, Made in China 2025 and the Future of American Industry, Centre for Strategic International Studies, 2019.
47. M. Fukuyama, "Society 5.0: Aiming for new human centered society," *Japan Spotlight*, vol. 27, no. 5, pp. 47–50, 2018.
48. W. L. Huan, "A data-driven decision support system for sustainable supplier evaluation in the Industry 5.0 era: A case study for medical equipment manufacturing," *Advanced Engineering Informatics*, vol. 56, pp. 101998, 2023.
49. J. Barata and I. Kayser, "Industry 5.0: Past, present, and near future," *Procedia Computer Science*, vol. 219, pp. 778–788, 2023.
50. Y. Lu, H. Zheng, S. Chand, W. Xia, Z. Liu, X. Xu and J. Bao, "Outlook on human-centric manufacturing towards Industry 5.0," *Journal of Manufacturing Systems*, vol. 62, pp. 612–627, 2022.
51. P. R. Madhikunta, Q. V. Pham, B. P. Deepa, K. Dev, T. R. Gadekallu and M. Liyanage, "Industry 5.0: A survey on enabling technologies and potential applications," *Journal of Industrial Information Integration*, vol. 26, p. 100257, 2021.
52. O. A. Elfar, C. K. Chang, H. Y. Leong, A. P. Peter, K. W. Chew and P. L. Show, "Prospects of Industry 5.0 in algae: Customization of production and new advance technology for clean bioenergy generation," *Energy Conversion and Management*, vol. 10, 2020.
53. M. Javaid and A. Haleem, "Critical components of Industry 5.0 towards a successful adoption in the field of manufacturing," *Journal of Industrial Integration and Management*, vol. 14, pp. 521–524, 2020.
54. F. Longo, A. Padovano and S. Umbrello, "Value-oriented and ethical technology engineering in Industry 5.0: A human-centric perspective for the design of the factory of the future," *Applied Sciences*, vol. 10, no. 12, p. 4182, 2020.
55. S. Nahavandi, "Industry 5.0: A human-centric solution," *Sustainability*, vol. 11, no. 16, pp. 13, 2019.
56. P. Sachsenmeier, "Industry 5.0: The relevance and implications of bionics and *synthetic* biology," *Engineering*, vol 2, no. 2, pp. 225–229, 2016.

2 Analysing the Research Landscape of Smart Cities and Artificial Intelligence

A Study Utilizing R Studio and VOSviewer

Nisha Kumari, Mukesh Kondala, and Atheer Abdullah Mohammed

2.1 INTRODUCTION

The adoption of several concepts, including sustainable cities, inclusive cities, and smart cities, has resulted in notable alterations of cities (Komninos et al., 2019). Population expansion, improvements in education, a rise in the need for municipal services, and most significantly, technology breakthroughs are the causes of these changes (Allam & Dhunny, 2019). Access to first-rate social amenities, as well as increased safety, connection, and transit, have all enhanced in smart cities. As a result, technologies like artificial intelligence (AI) are being incorporated into municipal management and growing in popularity (Daniels & Simpson, 2021).

While AI has been popular since the early 1950s, the adoption of its practical applications has lagged (Lande et al., 2020). According to a Google Trends investigation, a drop in AI can be seen between early 2004 and 2008 (Allam & Dhunny, 2019). However, it can be shown that from 2014 to 2017, the popularity of AI increased. Intriguingly, Allam (2018) pointed out that the idea of "Smart Cities" surged at the same time, highlighting a significant relationship between artificial intelligence and big data. A "smart everything" paradigm has emerged as an outcome of the rapid expansion of smart services powered by AI algorithms and the extensive usage of IoT. This paradigm encompasses different facets of our society and has a major impact on both our private and public life (Streitz, 2017). The advancement of smart cities is heavily reliant on information and communication technology (ICT). Automation, decentralization, democracy, and security are all traits shared by artificial intelligence and blockchain technology (Singh et al., 2020).

 DOI: 10.1201/9781003459835-2

A "smart city" is an urban system that aims to achieve efficiency and sustainability criteria in important domains and application areas, like administrative services, energy and environmental management, and transport (Bartolozzi et al., 2015; Pellicer et al., 2013). Whereas, AI contributes to the advancement of Smart Cities by improving operational efficiency, decision-making processes, security, data management, intelligent systems, and citizen engagement (Ahmad et al., 2022). Several crucial conditions must be addressed in order to build an AI-enabled smart city, guaranteeing security, scalability, and effectiveness (Mahor et al., 2023). These prerequisites include an environment that is resilient, interoperable, and flexible, energy sources and distribution, a decision-support system, behaviour monitoring, scalability, smart infrastructure, smart healthcare, and secure infrastructure (Mylonas et al., 2021; Singh et al., 2020).

A smart city generates enormous amounts of data, but without analysis to uncover important insights, the data is meaningless (Matheus et al., 2020). Within the context of smart city environments, AI is essential for processing and evaluating the data obtained from machine-to-machine communication. The deep learning and machine learning technologies facilitate streamlining the predictive and preventive decision-making process by understanding the intra and inter-system processes. AI assistance is crucial in analysing broad data of smart cities and facilitating the decision-making process (Ahad et al., 2020). It assists the smart city in various aspects, including resource allocation, city productivity, and improving living conditions. This technology helps create settings that are robust, creative, and resilient, and it also has the ability to extract data. The study examined "AI" and "Smart Cities" in synchrony with the aid of bibliometric analysis. The analysis examined a variety of published works and articles to ascertain the connections between them and any new developments in these two fields.

The bibliometric analysis explores the retrieved literature and represents the information about the performance of the retrieved literature and on emerging trends, authors, collaborations and study fields. This analysis also explores the various literature frameworks from the generated result. The analysis has been classified into two categories, that is, performance or descriptive analysis and content analysis. The performance analysis determines the various authors, affiliations and countries that contributed to the particular research field, and content analysis represents the evolution of the research field by highlighting the frequency and trends of the studied keywords (Kumari & Kondala, 2023). By utilizing the broad literature retrieved from the databases, this analysis provides an overview of the topic by examining and visualizing the data. This technique helps researchers study the evolution and the current trends of the specific research field.

2.2 METHODOLOGY

By examining the recent trends, authors, affiliations, and countries that have been doing research on that particular topic or keyword, the bibliometric analysis method works as a tool to study the evolution of the topic or keyword. Utilizing the SCOPUS database, the role of Artificial Intelligence and its connection with smart cities has been studied.

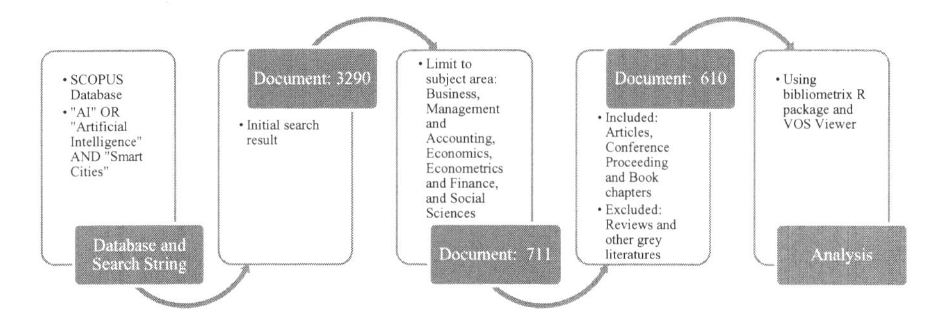

FIGURE 2.1 Data processing.

To retrieve and analyse the data using the bibliometric analysis method, the authors followed the following steps (see Figure 2.1). A search was done in the SCOPUS database using the terms "AI" OR "Artificial Intelligence" AND "Smart Cities" in titles, abstracts, and keywords to obtain bibliometric metadata about AI technology and smart cities. This method was used to make it easier to analyse pertinent scholarly works in this area. The researcher initially found 3,290 documents using the given search query. The most important documents showcasing the potential of AI in smart cities were extracted using inclusion and exclusion criteria. The initial phase of the search focused on the subjects of Finance and Econometrics, Economics, Business, Management and Accounting, and Social Sciences, ignoring data from other areas, including computer science, decision sciences, environmental sciences, etc. The database was then reduced to 711 documents. Only articles, conference papers, and book chapters were considered further, with reviews and other grey literature being excluded. As a result, the study incorporated 610 published research publications. The researcher chose not to reject publications based on publication year in order to better understand the topic's emergence and evolution across time, acknowledging that AI technology in smart cities has a historical context; however, adoption of AI practical applications lagged.

Descriptive statistics such as the average annual number of publications, the most productive authors, nations, institutions, and the most often cited author were calculated using the Bibliometrix R package. Additionally, content analysis was done to examine the co-occurrence, theme map, co-citation, and cooperation world map of the final dataset using the bibliometrix R package and VOSviewer in parallel.

2.3 RESULTS AND FINDINGS

2.3.1 Performance Analysis

2.3.1.1 Main Information

The main information about the data from the analysis is compiled in Table 2.1, which also includes relevant data on authors, countries, and authors' countries. The authorship data offers insightful information about the status of the writers and their cooperative efforts. As shown in Table 2.1, the publication of 610 documents was

TABLE 2.1
Summary of Data

Description	Results
Main Information about Data	
Sources (journals, books, etc.)	298
Documents	610
Annual growth rate (%)	0
Document average age	2.61
Average citations per doc	11.11
References	28,109
Document Contents	
Keywords plus (ID)	3,154
Author's keywords (DE)	2,027
Authors	
Authors	1,838
Authors of single-authored docs	110
Authors Collaboration	
Single-authored docs	114
Co-authors per doc	3.41
International co-authorships (%)	26.72
Affiliations	589
Countries	53
Document Types	
Article	257
Book chapter	106
Conference paper	247

created by 1,838 authors who were connected to 589 various institutions across 53 different countries and published in 298 sources.

2.3.1.2 Number of Publication per Year

The 610 documents in the sample are represented in Figure 2.2 by their annual publication and citation counts. In chronological sequence, the publishing trend shows an upward trend, culminating in 2022 with 128 publications. It shows this trend starting in 2011. The steady growth rate of 0% implies that there are plenty of prospects for more study in AI technology and smart cities. With a 2.61-year average age and an average of 11.11 citations per manuscript, it is clear that there is still work being done to explore new theories and concepts.

The first paper was published by Nadine M. Post in 2011, "Korea's Songdo IBD is model for sustainable, high-tech living", in this paper, he discussed Gale's U.Life pilot project to make the smart city work better; they used AI and next-generation technologies, which were all controlled by the master grid.

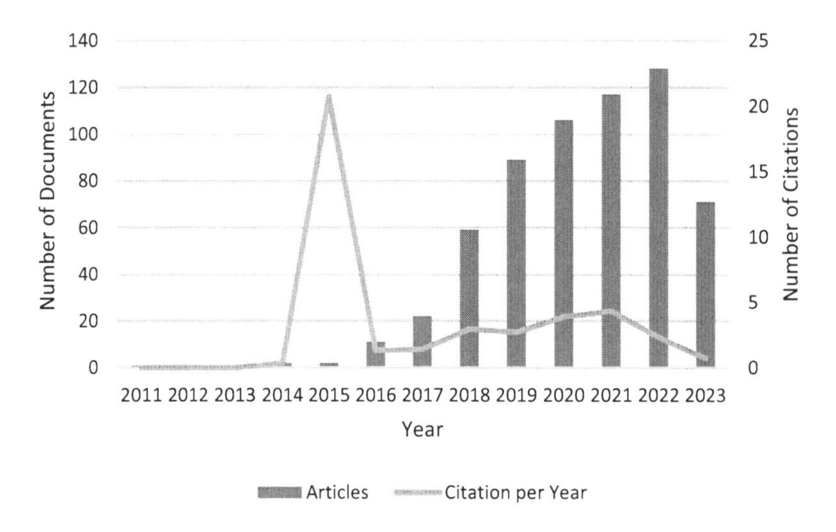

FIGURE 2.2 Annual scientific production and average citation per year.

Even though the average citation is more in 2015, the most cited paper is from 2019, "On Artificial Intelligence, Big Data and Smart Cities", with 454 citations, followed by the conference paper "Predicting short-term traffic flow by long short-term memory recurrent neural network" from 2015 publication with 350 citations. The difference is that in 2015, only two research works were published, whereas in 2019, 89 research works were published.

2.3.1.3 Most Relevant and Cited Journals

A total of 610 publications from 298 distinct sources were found in this analysis. The most prestigious journals in AI technology and smart cities research were identified using the Hirsch index (h-index). The h-index counts the number of scientific papers by a certain author or journal that have been referenced at least h times. It acts as a frequently used standard to evaluate the research output of authors and publications. The top 20 journals chosen based on their h-index are shown in Table 2.2. These journals are regarded as the most significant and pertinent research study sources. The one with the highest h-index of 17 among them, "*Sustainable Cities and Society*," has 27 publications and 1,409 citations, and its first publication in relation to AI and smart cities was made in 2018. Following closely is "*Sustainability (Switzerland)*", which has 44 publications, an h-index of 15, and 702 citations, and its first publication in 2016. Other prominent journals include "*Smart Cities*" (h-index 12, 496 citations, 23 publications, and its first publication in 2018) and "*2017 IEEE Smartworld Ubiquitous Intelligence and Computing, Advanced and Trusted Computed, Scalable Computing and Communications, Cloud and Big Data Computing, Internet of People and Smart City Innovation, Smartworld/SCALCOM/UIC/ATC/CBDCOM/IOP/SCI 2017- Conference Proceedings*" (h-index 8, 156 citations, 21 publications, and its first publication in 2018). The top 20 journals listed above can serve as important sources of information for researchers.

TABLE 2.2
Most Cited and Relevant Journals

Element	h_index	TC	NP	PY_start
Sustainable Cities and Society	17	1,409	27	2018
Sustainability (Switzerland)	15	702	44	2016
Smart Cities	12	496	23	2018
2017 IEEE Smartworld Ubiquitous Intelligence and Computing, Advanced and Trusted Computed, Scalable Computing and Communications, Cloud and Big Data Computing, Internet Of People and Smart City Innovation, Smartworld/Scalcom/Uic/Atc/Cbdcom/Iop/Sci 2017 – Conference Proceedings	8	156	21	2018
2017 International Smart Cities Conference, ISC2 2017	3	53	3	2017
Advanced Sciences and Technologies for Security Applications	3	30	13	2019
Big Data And Cognitive Computing	3	24	5	2021
IEEE 2nd International Smart Cities Conference: Improving the Citizens' Quality Of Life, ISC2 2016 – Proceedings	3	51	4	2016
Internet of Things (Netherlands)	3	57	3	2020
Proceedings – 2016 International Conference on Intelligent Transportation, Big Data and Smart City, ICITBS 2016	3	40	6	2017
Proceedings – 2019 IEEE Smartworld, Ubiquitous Intelligence and Computing, Advanced and Trusted Computing, Scalable Computing and Communications, Internet of People and Smart City Innovation, Smartworld/UIC/ATC/SCALCOM/IOP/SCI 2019	3	33	18	2019
Safety Science	3	94	3	2021
Smart Cities: Issues and Challenges Mapping Political, Social and Economic Risks and Threats	3	34	3	2019
Technological Forecasting and Social Change	3	122	3	2020
Technology in Society	3	92	3	2020
2016 IEEE International Conference on Smart Computing, SMARTCOMP 2016	2	32	2	2016
2018 IEEE International Smart Cities Conference, ISC2 2018	2	15	5	2019
2020 IEEE International Smart Cities Conference, ISC2 2020	2	7	2	2020
2021 IEEE European Technology and Engineering Management Summit, E-TEMS 2021 – Conference Proceedings	2	9	2	2021
5th IEEE International Smart Cities Conference, ISC2 2019	2	11	3	2019

2.3.1.4 Prominent Authors

The top 20 writers in the study are listed in Table 2.3, along with their Total Citations (TC), Number of Publications (NP), and PY-start scores. According to their h-index, ALLAM Z, YIGITCANLAR T, and MEHMOOD R are the top three writers. ALLAM Z has 7 publications, 7 h-index, and 591 citations, YIGITCANLAR T has 5 publications with 5 h-index and 287 citations, and MEHMOOD R (with 4 publications, 4 h-index, and 136 citations). ALLAM Z and YIGITCANLAR T published their first papers in their respective fields of study in 2019 and 2020, respectively, while MEHMOOD R published their first papers in AI and smart cities in 2021.

TABLE 2.3
Most Cited and Relevant Authors

Element	h_index	TC	NP	PY_start
Allam Z	7	591	7	2019
Yigitcanlar T	5	287	5	2020
Mehmood R	4	136	4	2021
Park S	4	103	4	2019
Al-Turjman F	3	123	3	2020
Bibri SE	3	74	3	2022
Chen Y	3	24	4	2020
Cugurullo F	3	220	3	2020
Gehlot A	3	49	4	2021
Li Y	3	47	4	2017
Nesi P	3	86	3	2015
Sharma A	3	68	3	2018
Singh R	3	51	5	2021
Singh S	3	248	4	2017
Abdel-Basset M	2	30	2	2019
Adamec V	2	4	2	2017
Ahmed S	2	20	2	2019
Alswedani S	2	25	2	2022
Anastasiu DC	2	34	2	2018
Antoniou J	2	18	2	2019

2.3.1.5 Most Contributed Countries

Contributions from 53 different countries have been made to the articles published on the subject of AI technology and smart cities. In order to emphasize the most active nations in this field of study, Figure 2.3 shows the distribution of Multiple Country Publications (MCP) and Single Country Publications (SCP). The top two nations in terms of SCP are China and India, with the United States, the United Kingdom, Saudi Arabia, and Korea rounding out the top five. On the other side, when it comes to MCP, Australia and Saudi Arabia are at the top, closely followed by India, Korea, and China. The top 10 countries for research on AI and smart cities are shown in Table 2.4, ranked according to the number of publications (f) gleaned from an analysis of each country's scientific production. With 263 and 244 publications, respectively, China and India stand out as the top two producing countries of scientific literature. Following closely after the United States with 226 publications are Italy and the United Kingdom, with 123 and 116 publications, respectively.

2.3.2 Content Analysis

2.3.2.1 Authors Co-citation Network

The degree of the association between the two papers is revealed through co-citation analysis, which reveals the frequency with which they are referenced together.

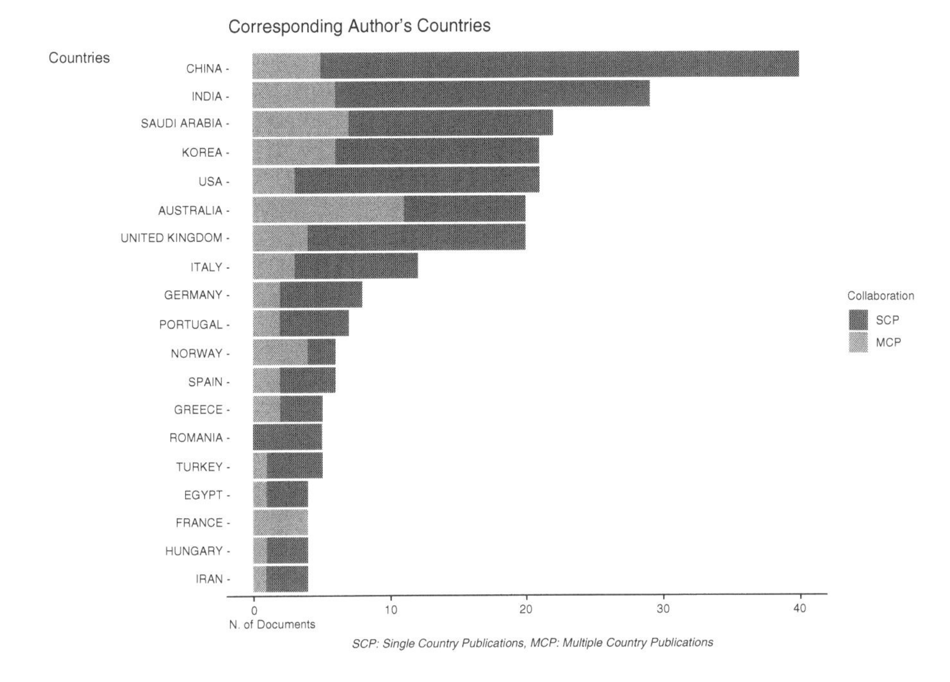

FIGURE 2.3 Contribution of countries in terms of publication.

TABLE 2.4
Top 10 Contributed Country's Scientific Production

Region	Frequency
China	263
India	244
United States	226
Italy	123
United Kingdom	116
Saudi Arabia	108
South Korea	74
Germany	68
Australia	60
Brazil	45

As co-citation frequency rises, so does the co-citation strength of these two articles. Researchers can progressively determine the most prominent authors in a specific field of study in an author co-citation network based on the citations they exchange with other authors.

Using VOSviewer, co-citations between authors were examined and displayed. Finding well-known authors connected to the study area was made easier with the use

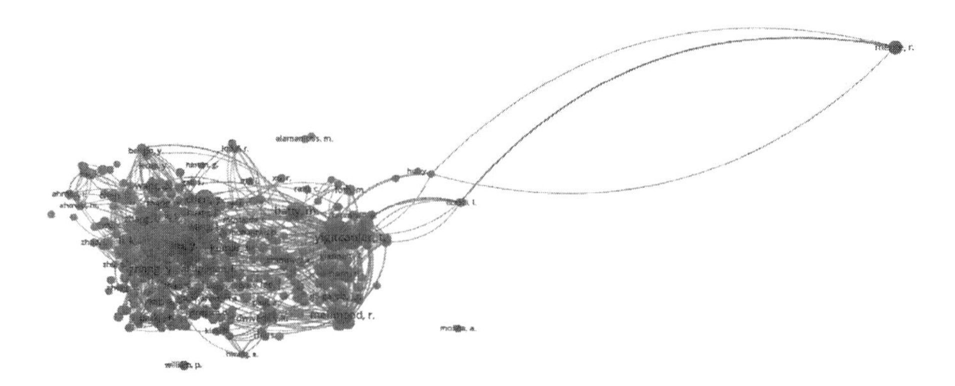

FIGURE 2.4 Author's co-citation network.

of this tool. Using VOSviewer, the authors were chosen, and the following co-citation analysis allowed for a thorough grasp of scholarly relationships and subject matter competence. A total of 42,293 authors were on the list; among them, only 402 had at least 15 citations. These 402 writers were shown as nodes in 6 clusters inside the co-citation network (Figure 2.4).

The top three authors in each cluster were determined based on the number of citations each author received. Wang, X (110 citations and 6,574 link strength), Wang, Y (109 citations and 5,926 link strength), and Wang, J (101 citations and 6,127 link strength) were the most notable authors in the first cluster (red). Zhang, Y (123 citations and 7,306 link strength), Liu, Y (119 citations and 5,808 link strength), and Kumar, S were all part of the second cluster (green) (74 citations and 2,050 link strength). Batty, M (93 citations and 3,691 link strength), Nam, T (61 citations and 2,492 link strength), and Komninos, N (54 citations and 2,153 link strength) were prominent in the third cluster (navy blue). Yigitcanlar, T (162 citations and 10,461 links), Allam, Z (105 citations and 4,963 links), and Mehmood, R (95 citations and 7,026 link strength) made up the fourth cluster (yellow). Wang, G (28 citations and 1,452 link strength) was highlighted in the fifth cluster (purple), along with Ullah, Z (22 citations and 1,561 link strength) and Mohanty, SP (22 citations and 1,154 link strength). Last but not least, Shen, X (18 citations and 778 link strength), Molina, A (17 citations and 407 link strength), and Kim, D (16 citations and 410 link strength) made up the sixth cluster (blue).

2.3.2.2 Keyword Co-occurrence Network Analysis

The keyword co-occurrence network analysis, as depicted in Figure 2.5, is performed using the R program language. This network analysis will provide a greater knowledge of the interaction dynamics between AI technologies and smart cities. By displaying/visualizations of correlations and patterns between keywords, a co-occurrence network offers valuable insights into research trends and themes in the subject. According to Kumari and Kondala (2023), when two terms occur together in publications, they are displayed as linkages between two nodes in a network of keywords, where each word is represented as a node. The frequency with which two terms co-occur in distinct publications serves as a proxy for the strength of the relationship between them.

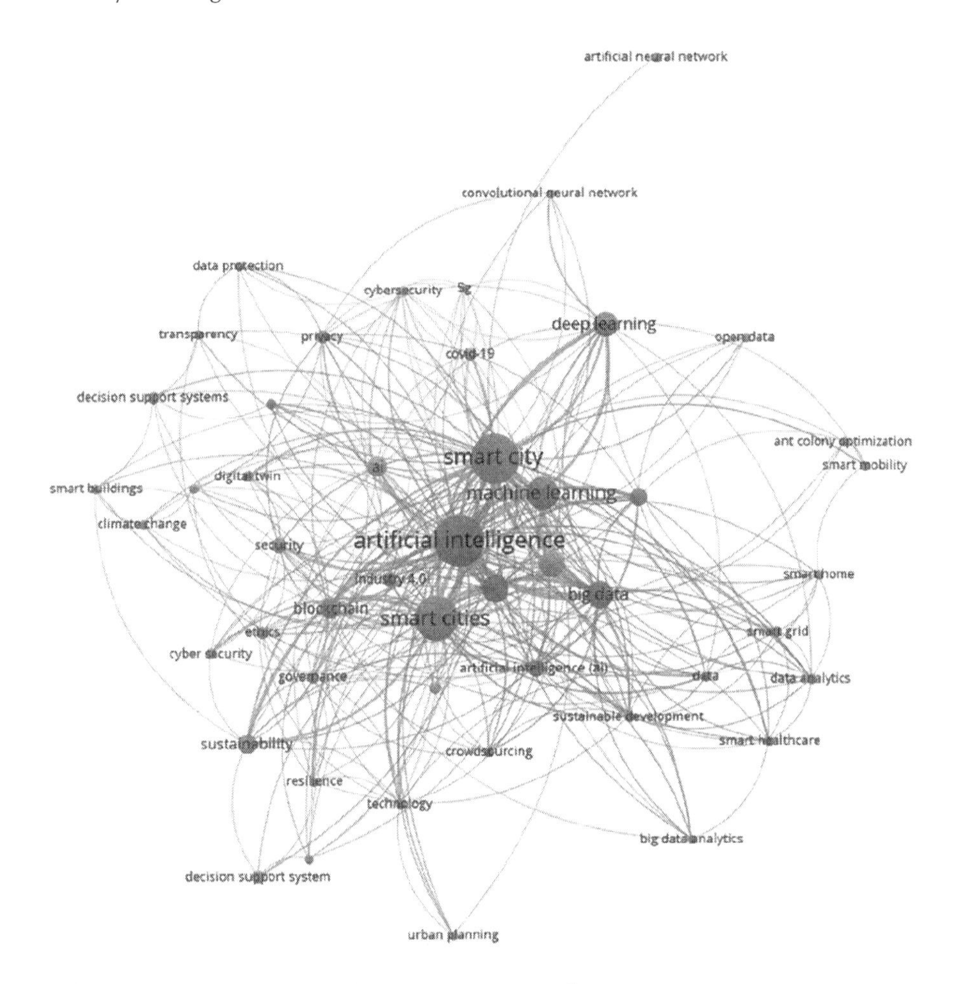

FIGURE 2.5 Author's keyword co-occurrence network.

From the keyword co-occurrence analysis, the seven different clusters were found in seven different colour representations, shown in Figure 2.5. The first biggest cluster is highlighted in red which includes prominent keywords like cyber security, decisions support systems, governance, urban planning, resiliency, and technology, which are the important elements for accomplishing the objectives of sustainable smart city. The second cluster, which is highlighted in green colour, and the keyword strengths are related to technology and various smart elements impacting the creation of smart cities like artificial intelligence (AI), the Internet of Things (IoT), data analytics, smart grid, smart healthcare, and smart transportation.

The following clusters, from third to seventh, are shown in different colours, covering various concepts and elements related to developing a smart and sustainable world. These key components promote Blockchain (BC), ML, cloud computing, deep learning, convolutional neural networks, decision-support systems, big data analytics, crowdsourcing, metaverse, transparency, innovation, and Industry 4.0, some of the clusters that comprise this list.

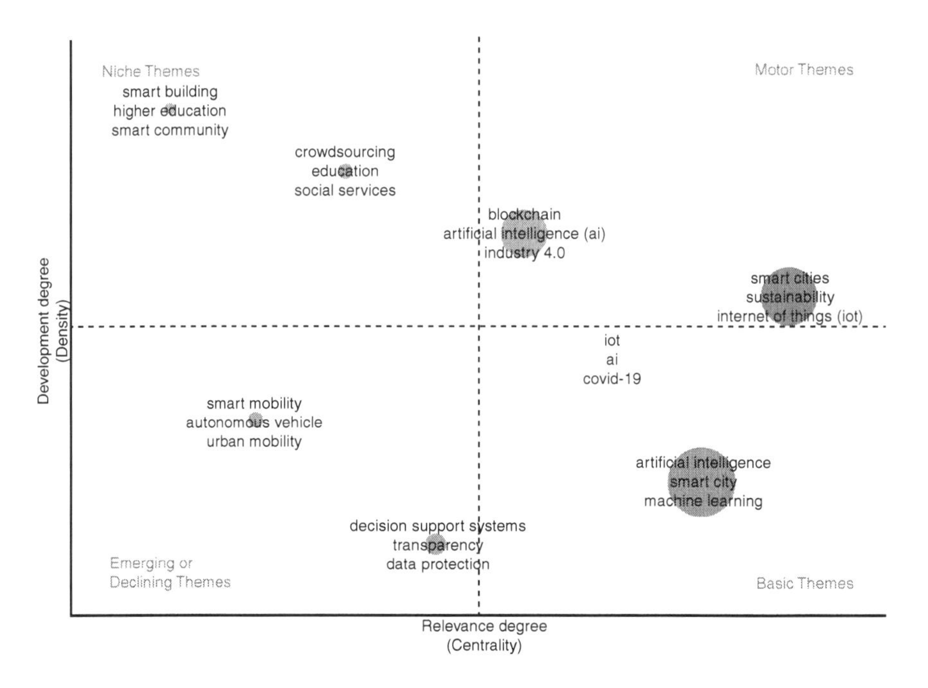

FIGURE 2.6 Thematic map of smart cities and artificial intelligence.

2.3.2.3 Thematic Map

The thematic map gives a substantial advantage by allowing the research intensity based on the various levels of density and prominence. A theme's density shows how it has changed and gained prominence over time, whereas a theme's centrality is defined by how closely it is related to other themes. The most significant and influential themes in the subject of study can be identified according to this capacity, which offers insightful information about the distribution and connections between research topics.

As shown in Figure 2.6, the fundamental components of smart cities, such as crowdsourcing, smart communities, and higher education, are classed as niche topics. On the other hand, aspects related to mobility and transportation, such as autonomous vehicles and smart mobility, are noted as either emerging or declining topics. Transparency, data protection, and decision-support systems, however, are also categorized as emerging or decreasing themes even though they have a high level of relevance with other themes. Smart cities, AI, and Machine Learning (ML), which stand as the fundamental foundations, are the study's basic themes. Additionally, the IoT and AI and BC in relation to Industry 4.0 are acknowledged as motor themes, indicating their prominent and influential positions within the research environment.

2.3.2.4 Countries Collaboration Worldmap

On a map of the world, Figure 2.7 shows global patterns in research collaboration. China and the United States have the most productive cooperation (f=7), followed

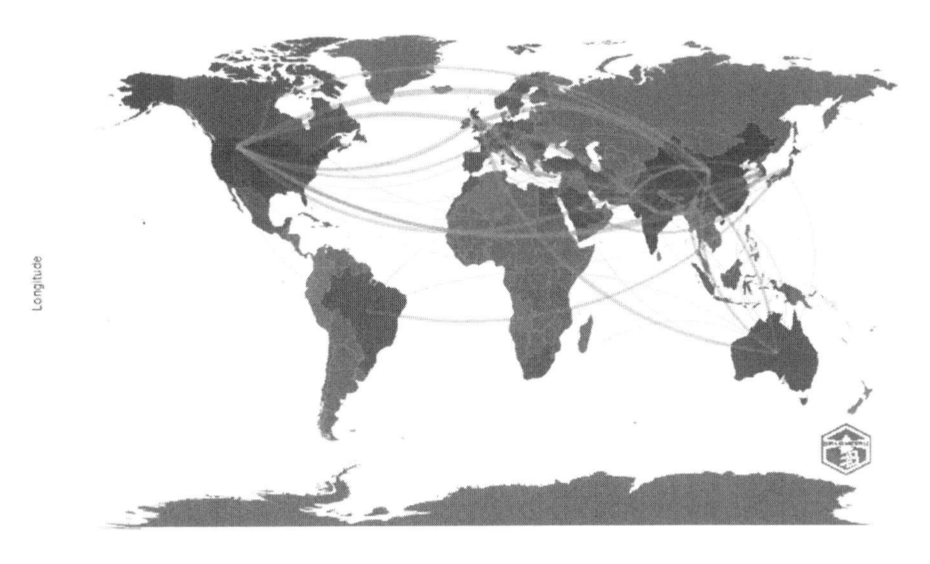

FIGURE 2.7 Country's collaboration world map.

TABLE 2.5
Top 10 Country's Collaboration

From	To	Frequency
China	United States	7
China	United Kingdom	5
India	United States	5
India	Korea	4
India	Saudi Arabia	4
China	Hong Kong	4
United Kingdom	Australia	4
United Kingdom	Korea	4
United States	Japan	4
United States	Korea	4
United States	Saudi Arabia	4
United States	United Kingdom	4
Saudi Arabia	Pakistan	4

by China and the United Kingdom (f=5) and India and the United States (f=5). The top 10 collaborations are shown in Table 2.5, along with the equivalent number of works created. These results indicate that worldwide collaborative networks in the study of AI technologies and smart cities are not considerably influenced by geographic or language proximity.

2.4 DISCUSSION

This study's bibliometric analysis offers important new insights into the field of business, management and accounting, finance, econometrics, economics, and social sciences research on artificial intelligence and smart cities. The study's key finding is the explosive expansion of research into smart cities and artificial intelligence. The rise in publications over time, with a peak in 2022, is a reflection of the significance and interest in this subject among academics. The long average age of the publications and the large number of citations per paper suggest that this field's researchers are constantly pursuing novel concepts and expanding the frontiers of knowledge.

The evaluation of renowned publications and authors yields insightful data about the most significant field contributions. Based on their h-index and citation counts, journals like "*Sustainable Cities and Society*" and "*Sustainability (Switzerland)*" have emerged as the most respected sources of research on AI technology and smart cities. Similarly, writers with high citation counts and h-index scores, such as Allam Z, Yigitcanlar T, and Mehmood R, are acknowledged for their substantial contributions to this field of study. The Allam Z is the lead author in five of his total publications. In his research, he addressed about the concept of smart cities emerging with new technologies like AI and big data. He also investigated the essentially liveable cities. modernizing the use of technology in urban health to improve the quality of life in smart cities, all the while taking into account the potential and obstacles for achieving environmental, social, and economic sustainability in metropolitan settings. In order to accomplish the objectives of smart cities, this entails designing a three-dimensional framework that integrates virtual environments, the metaverse, and artificial neural networks in harmony with policy choices and urban governance (Allam & Dhunny 2019; Allam, 2019; Allam et al., 2019, 2022; Bibri & Allam, 2022). When it comes to country contributions, China is the leading contributor with 263 publications, followed by India and the United States with 244 and 226 publications, respectively. China and India are notable countries when it comes to SCP, while Australia and Saudi Arabia are the most prominent countries when it comes to MCP.

Additionally, an analysis of author co-citations reveals that Wang X and Wang Y are the most often cited authors. While Wang X's research examines various aspects of smart cities, including smart manufacturing, real-time data traffic management, data integration, blockchain for smart cities, etc. (Altulyan et al., 2020; Ning et al., 2019; Suvarna et al., 2020; Wang et al., 2016), Wang Y's research focuses mainly on the effects of smart cities, including CO_2 emission reduction, energy savings, a sustainable low-carbon future, latency optimization, etc. (Guo et al., 2022; Tang et al., 2020; Wang et al., 2019). The thematic map displayed significant term clusters that highlighted the main areas of research in the domain. Urban planning, technology, governance, and sustainable development issues have all become significant contributors to the goals of smart cities. On the other hand, new or waning themes like mobility and transportation draw attention to how this area of research is constantly changing.

The analysis shows that there is much use of AI, big data, and machine learning are the best technologies for making smart cities better and sustainable. There are many countries involved in the contributions of these technologies in research, funding, and implementation. Some of them are China, India, the United Kingdom, and the United States; many other countries are also contributing to the knowledge of smart cities. It also tells us that the researchers from these countries are cooperating with each other regarding knowledge sharing and showing some impacts in progression.

2.4.1 Theoretical Implication

The theoretical Implications of the study on the evolution of AI and across time cover significant topics in the field, promising authors, and international collaborations in the study area. The study can unveil the truth about the field's state of the art and developments happening along with research. The study can expand the investigations of the usage of AI in a sustainable manner and make cities more sustainable and also lay the foundations for the implementation phases. There are lot of insights from the study on how AI influencing the concepts innovative and developments in urban planning. This AI will play a bigger role in influencing the cities to develop in the future, and as also advancing in the technologies, cities should get more sophisticated, effective, and eco-friendly. The theme map and keyword analysis offer insightful information about the relationships and importance of different variables. This aids in comprehending the developing themes and focus points in the study of smart cities and artificial intelligence.

2.4.2 Practical Implications

There are several useful advantages to researching AI technology and smart cities. It assists scholars and decision-makers in locating reliable resources and collaborators by showcasing the top papers, notable individuals, and countries in this area. Prioritizing research for efficient urban planning and policymaking also benefits from understanding current trends. Developing comprehensive and flexible solutions for smart cities requires understanding the close relationships between AI, related technologies, and smart cities. Comprehending the collaborative dynamics among researchers fosters worldwide cooperation and information exchange, thereby supporting inventive and eco-friendly smart city advancement. We can obtain important insights for timely decision-making and investment strategy development to maximize the revolutionary potential of artificial intelligence in urban development by analysing publishing trends. These insights provide stakeholders with the necessary tools to create sustainable and technologically advanced smart city designs.

2.5 CONCLUSION

Urbanization and globalization present problems for cities, and robust infrastructure is essential for them to become sustainable smart cities. Interoperable, adaptable, and flexible communication technologies are essential for effective communication in these cities. In smart cities, adaptive decision-support systems are essential for juggling the demands of industry, the environment, and the economy. Simulated behaviour analysis reveals intricate interaction patterns. Energy delivery and sources must be dependable, especially for renewable resources. Intelligent systems make coordination and cooperation easier. Due to the large number of hardware and software components, scalability is crucial, necessitating distributed IoT-blockchain cloud infrastructures. ICT integration and sophisticated analytics are advantageous for smart healthcare. Networks and data are protected by secure infrastructure that includes monitoring, authentication, and authorization in sustainable smart cities.

The findings emphasize the crucial part that AI technology will play in creating the smart cities of the future. Urban planning, transportation, healthcare, and energy management can all be dramatically improved by integrating AI, machine learning, and other cutting-edge technology. These technologies have the potential to lead to more effective and environmentally friendly city operations, better quality of life for locals, and efficient policymaking. Additionally, the analysis shows that although there is a historical framework for the adoption of AI technology in smart cities, actual applications have lagged. This finding emphasizes the value of ongoing research and innovation to get around obstacles and fully utilize AI's potential to revolutionize urban areas.

2.5.1 Limitations and Scope for Future Research

The analysis is based on information from the SCOPUS database, which could not include all pertinent academic papers. Additionally, the results can have been affected by the choice of keywords and inclusion standards. Future studies can think about combining information from other databases and using a wider range of keywords and inclusion criteria in order to acquire a more complete understanding of the research environment. A deeper understanding of the context and impact of AI in smart cities can also be gained by using qualitative methodologies. In order to promote the creation of evidence-based policies for sustainable and technologically advanced urban development, it would be beneficial to explore multidisciplinary links and longitudinal patterns.

REFERENCES

Ahad, M. A., Paiva, S., Tripathi, G., & Feroz, N. (2020). Enabling technologies and sustainable smart cities. *Sustainable Cities and Society*, *61*, 102301.

Ahmad, K., Maabreh, M., Ghaly, M., Khan, K., Qadir, J., & Al-Fuqaha, A. (2022). Developing future human-centered smart cities: Critical analysis of smart city security, data management, and ethical challenges. *Computer Science Review*, *43*, 100452.

Allam, Z. (2018). Contextualising the smart city for sustainability and inclusivity. *New Design Ideas*, *2*(2), 124–127.

Allam, Z. (2019). Achieving neuroplasticity in artificial neural networks through smart cities. *Smart Cities*, *2*(2), 118–134.

Allam, Z., & Dhunny, Z. A. (2019). On big data, artificial intelligence and smart cities. *Cities*, *89*, 80–91.

Allam, Z., Sharifi, A., Bibri, S. E., Jones, D. S., & Krogstie, J. (2022). The metaverse as a virtual form of smart cities: Opportunities and challenges for environmental, economic, and social sustainability in urban futures. *Smart Cities*, *5*(3), 771–801.

Allam, Z., Tegally, H., & Thondoo, M. (2019). Redefining the use of big data in urban health for increased liveability in smart cities. *Smart Cities*, *2*(2), 259–268.

Altulyan, M., Yao, L., Kanhere, S. S., Wang, X., & Huang, C. (2020). A unified framework for data integrity protection in people-centric smart cities. *Multimedia Tools and Applications*, *79*, 4989–5002.

Bartolozzi, M., Bellini, P., Nesi, P., Pantaleo, G., & Santi, L. (2015). A smart decision support system for smart city. In *2015 IEEE International Conference on Smart City/SocialCom/SustainCom (SmartCity)* (pp. 117–122). IEEE, Chengdu, Sichuan, China.

Bibri, S. E., & Allam, Z. (2022). The Metaverse as a virtual form of data-driven smart cities: The ethics of the hyper-connectivity, datafication, algorithmization, and platformization of urban society. *Computational Urban Science*, *2*(1), 22.

Daniels, C., & Simpson, M. (2021). *City for the Future: Resilient, Equitable, and Regenerative*, Environmental Science, Engineering, Sociology.

Guo, Q., Wang, Y., & Dong, X. (2022). Effects of smart city construction on energy saving and CO2 emission reduction: Evidence from China. *Applied Energy*, *313*, 118879.

Komninos, N., Kakderi, C., Panori, A., & Tsarchopoulos, P. (2019). Smart city planning from an evolutionary perspective. *Journal of Urban Technology*, *26*(2), 3–20.

Kumari, N., & Kondala, M. (2023). Role of soft power in tourism development: A bibliometric analysis of the past decade. In *Global Perspectives on Soft Power Management in Business* (pp. 245–260). IGI Global, India, USA.

Lande, Dmytro and Strashnoy, Leonard and Balagura, Irina, Directed correlation network of concepts determined by the dynamics of publications (September 8, 2020). Available at SSRN: https://ssrn.com/abstract=3688659 or http://dx.doi.org/10.2139/ssrn.3688659.

Mahor, V., Rawat, R., Kumar, A., Garg, B., & Pachlasiya, K. (2023). IoT and artificial intelligence techniques for public safety and security. In *Smart Urban Computing Applications* (pp. 111–126). River Publishers, India, Denmark.

Matheus, R., Janssen, M., & Maheshwari, D. (2020). Data science empowering the public: Data-driven dashboards for transparent and accountable decision-making in smart cities. *Government Information Quarterly*, *37*(3), 101284.

Mylonas, G., Kalogeras, A., Kalogeras, G., Anagnostopoulos, C., Alexakos, C., & Muñoz, L. (2021). Digital twins from smart manufacturing to smart cities: A survey. *IEEE Access*, *9*, 143222–143249.

Ning, Z., Huang, J., & Wang, X. (2019). Vehicular fog computing: Enabling real-time traffic management for smart cities. *IEEE Wireless Communications*, *26*(1), 87–93.

Pellicer, S., Santa, G., Bleda, A. L., Maestre, R., Jara, A. J., & Skarmeta, A. G. (2013). A global perspective of smart cities: A survey. In *2013 Seventh International Conference on Innovative Mobile and Internet Services in Ubiquitous Computing* (pp. 439–444). IEEE, Taichung, Taiwan.

Singh, S., Sharma, P. K., Yoon, B., Shojafar, M., Cho, G. H., & Ra, I. H. (2020). Convergence of blockchain and artificial intelligence in IoT network for the sustainable smart city. *Sustainable Cities and Society*, *63*, 102364.

Streitz, N. (2017). Re-defining the "smart everything" paradigm: Towards reconciling humans and technology. In *Proceedings of the International Conference on ICT, Society and Human Beings,* Germany.

Suvarna, M., Büth, L., Hejny, J., Mennenga, M., Li, J., Ng, Y. T., … & Wang, X. (2020). Smart manufacturing for smart cities: Overview, insights, and future directions. *Advanced Intelligent Systems*, *2*(10), 2000043.

Tang, C., Wei, X., Zhu, C., Wang, Y., & Jia, W. (2020). Mobile vehicles as fog nodes for latency optimization in smart cities. *IEEE Transactions on Vehicular Technology*, *69*(9), 9364–9375.

Wang, Y., Ren, H., Dong, L., Park, H. S., Zhang, Y., & Xu, Y. (2019). Smart solutions shape for sustainable low-carbon future: A review on smart cities and industrial parks in China. *Technological Forecasting and Social Change*, *144*, 103–117.

Wang, S., Wang, X., Wang, Z. L., & Yang, Y. (2016). Efficient scavenging of solar and wind energies in a smart city. *ACS Nano*, *10*(6), 5696–5700.

3 Enhancing Digital Accessibility for AI-Powered Interfaces in Smart Cities

Parampreet Singh

3.1 INTRODUCTION

We share this world with 1.3 billion people who have a disability (WHO's disability report [1]), which represents 16% of the world's population, or 1 in 6. Now, imagine what a smart city solution would be like when a person with no vision cannot interact with an AI-powered web application to make hotel, car, or bus reservations or a person with limited hand movement is unable to operate devices (self-checkout kiosks) that require sophisticated muscle movements. Can that city be a SMART CITY?

The design and development of web applications/digital interfaces with accessibility issues affects not only the 41 million Americans with disabilities but also has broader implications. Even with a robust U.S. Labor market, individuals with disabilities face noticeable underemployment. As per the annual report on People with Disabilities in America 2023 [2], only 38.4% of working-age Americans with disabilities (between ages 18 and 64) were part of the workforce, compared to 75.8% of Americans without disabilities in 2021. A report by the American Institutes for Research in 2018 titled "The purchasing power of working-age adults with Disabilities" [3] revealed that the after-tax income for working-aged individuals with disabilities is around $490 billion, comparable to market segments like African Americans ($501 billion) and Hispanics ($582 billion). On average, individuals with disabilities have a disposable income of approximately $23,300 after taxes, which is 35% lower than those without disabilities. People with disabilities form a significant consumer market, identified as the third-largest segment in the United States by the U.S. Office of Disability Employment Policy (ODEP) [4]. When considering their families, caregivers, and others who prioritize inclusivity, this market size doubles.

These reports indicate that Digital Inclusion can play a vital role in providing equitable opportunities for people with disabilities to work, earn a living, and contribute to the nation's progress.

DOI: 10.1201/9781003459835-3

3.1.1 Background

Sustainable Development Goals (SDGs) [5] set by the United Nations for 2030 clearly outline the focus on social inclusion, aiming to end poverty, protect the planet, and ensure that all people (irrespective of their abilities) enjoy peace and prosperity, and the concept of smart cities is gaining a lot of attention as urban areas strive to leverage technology for improved efficiency, sustainability, and quality of life. Integration of modern technologies, such as Internet of Things (IoT) devices and Artificial Intelligence (AI) systems, plays a key role in creating interconnected and data-driven environments. However, the thing that is commonly overlooked is to make them inclusive, ensuring that people with disabilities have equitable access to them.

Digital accessibility is a fundamental human right and is the key to building inclusive smart cities. It involves providing equitable access to the internet, web applications, and digital interfaces such as ATM machines, empowering them to participate fully in the inclusive ecosystems that cater to the need of diverse citizens.

Smart cities encompass crucial aspects of life, including education, employment, government, healthcare, and various other essential elements. It is critical that AI and IoT-powered websites and interfaces are accessible to provide equitable access to people with diverse abilities. By prioritizing accessibility, smart cities enable people with disabilities to lead independent lives. Accessible infrastructure, assistive technologies, and supportive services empower them to navigate the city, communicate effectively, and engage in various activities without unnecessary reliance on others.

The significance of accessibility in smart cities cannot be overstated. By ensuring equitable access and opportunities for people with disabilities, smart cities can create inclusive and sustainable environments that benefit the entire community. Embracing accessibility not only aligns with legal obligations but also reflects a city's commitment to equity, social responsibility, and the well-being of its residents and visitors.

3.1.2 Objectives

This chapter focuses on the following objectives:

1. Define and understand disability, digital accessibility, governing laws, assistive technologies, and limitations of existing solutions.
2. An overview of the Web Content Accessibility Guidelines (WCAG) [6] and Web Accessibility Initiative – Accessible Rich Internet Applications (WAI-ARIA) [7].
3. Explore and highlight the significance of a human-centred design approach for AI-powered interfaces.
4. Evaluate the current state of accessibility for existing travel booking websites for error density and top error types.
5. Provide recommendations for design and development of accessible AI and IoT-based interfaces in smart cities.

3.2 DISABILITY, DIGITAL ACCESSIBILITY, LAWS, AND ASSISTIVE TECHNOLOGIES

Defining disability is a complex phenomenon and it can be best defined when viewed from different lenses. Theoretical model of disability is one approach that helps us define disability, discusses strengths and weaknesses of different types of models.

3.2.1 Theoretical Models of Disability

The assumptions about people with disabilities have a profound impact on how digital technologies are planned, designed, and developed. These assumptions shapes government policies, employment opportunities, and advancements of smart cities.

There are many theoretical models such as the Medical, Social, Biopsychosocial, Economic, Functional, Social Identity or Cultural Affiliation, and Charity models that define "Disability" from their respective lenses. However, for this chapter, it's vital to define and understand disability from two theoretical models viz. Medical and Social.

3.2.1.1 Medical Model

A medically diagnosed genetic disorder, disease, or other health conditions are viewed as problematic and are considered as biological issues that can impact the quality of life [8]. The Centers for Disease Control and Prevention (CDC) [9] defines disability as a condition of the body or mind that makes certain activities more challenging (activity limitation) and hinders interaction with the surrounding world (participation restrictions) (Table 3.1).

Various types of disabilities that can impact individuals, including:

a. Vision
b. Movement
c. Thinking
d. Remembering

TABLE 3.1
Strengths and Weaknesses of the Medical Model

Strengths	Weaknesses
• In a medical setting, a well-defined set of biological criteria to diagnose a person's condition helps medical professionals make life saving decisions • Clearly defined set of criteria helps governments decide who should receive government assistance	• It focuses exclusively on the biology of a person, overlooking the impact of design decisions in our social environment • By labelling and stigmatizing individuals with disabilities, it creates a sense of "otherness" and perpetuates the notion that they are somehow inferior to the rest of the population. This not only affects the individuals themselves but also has a profound impact on the overall fabric of society

e. Learning
f. Communicating
g. Hearing
h. Mental health

3.2.1.2 Social Model

Social model [10] diverges from the dominant medical model and is a direct response to the medical model, wherein it points out that society creates disabling conditions by not planning, designing, and developing for people with disabilities. Social model redefines disability as the restrictions caused by society by not giving equitable social and structural access. In short, disability is an avoidable condition resulting from poor design.

Social model of disability guides and provides a path for professionals to make accessible websites, buildings, consumer products, and transportation systems, ensuring that anything that is created is not disabling for people with diverse physical abilities (Table 3.2).

Following disabling environments are the problem:

a. Web designers
b. Architects
c. Product designers
d. Public planners
e. Higher expectations

Disability is a spectrum rather than a simple binary characteristic of a person, and disability can be characterized as follows:

1. **Permanent**: A defining characteristic of a person's body.
2. **Temporary**: An injury, sickness, or short-term impairment.
3. **Situational**: A condition or context that limits a person's ability, for example a woman holding her newborn baby.

TABLE 3.2
Strengths and Weaknesses of the Medical Model

Strengths	Weaknesses
• It empowers and removes the stigma associated with physical impairments and expects that society will build inclusive environments for people with disabilities to thrive • It also empowers designers of both virtual and physical environments to include diverse personas, resulting in endless opportunities, and offering higher quality of life	• Some disability advocates argue that adopting social models of disability can help shift the focus away from physical limitations and instead promote acceptance and embracing of one's disability. From an emotional and psychological perspective, this can lead to a healthier mindset and overall well-being

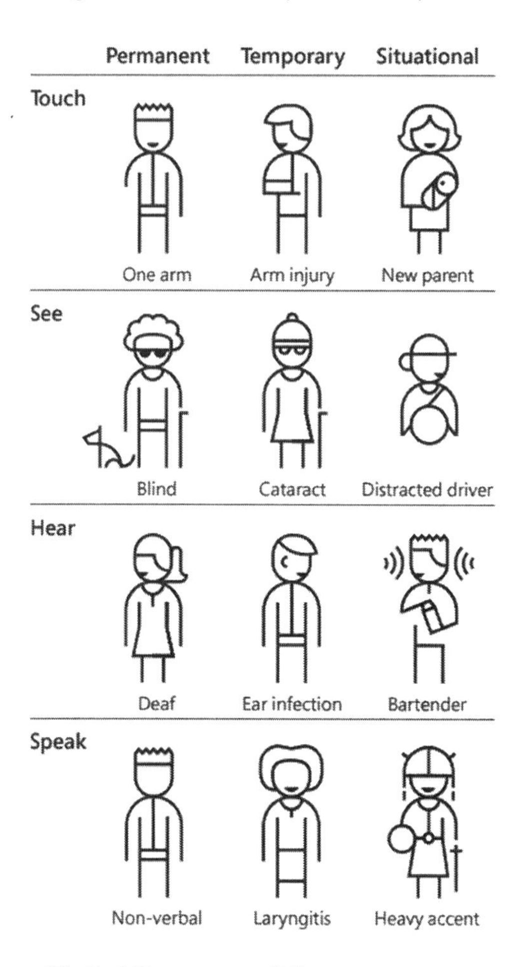

FIGURE 3.1 Microsoft's disability spectrum [11].

Figure 3.1 is Microsoft's tabular illustration of disability spectrum with four rows and four columns (human sense, permanent, temporary, and situational); rows details are as follows:

1. Row 1 has "Touch" sense with one arm under permanent, arm injury under temporary, and new parent under situational disability.
2. Row 2 has "See" sense with blind under permanent, Cataract under temporary, and distracted driver under situational disability.
3. Row 3 has "Hear" sense with deaf under permanent, ear infection under temporary, and bartender under situational disability.
4. Row 4 has "Speak" sense with non-verbal under permanent, Laryngitis under temporary, and heavy accent under situational disability.

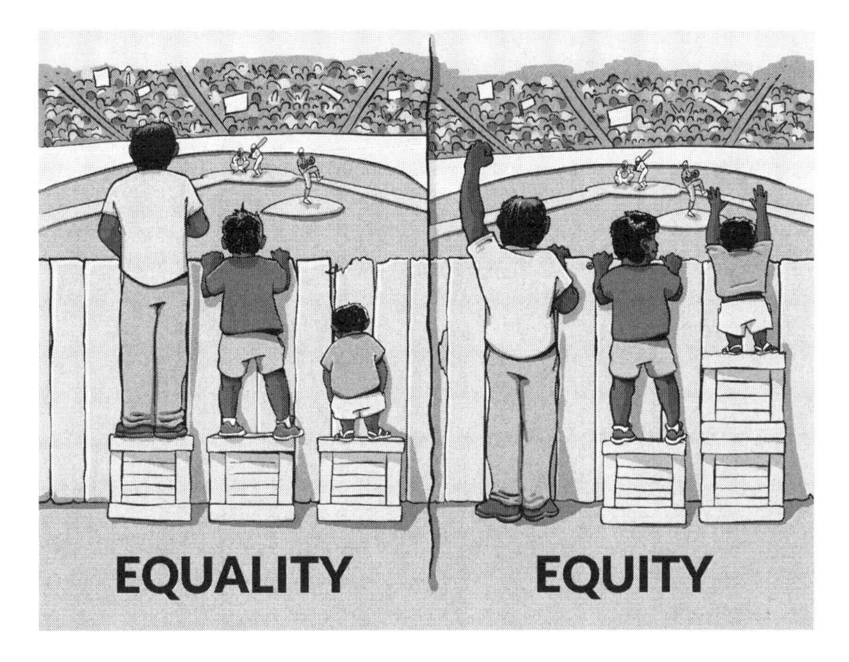

FIGURE 3.2 Interaction institute for social change | artist: angus Maguire [13].

3.2.2 Digital Accessibility and Its Benefits

According to W3.org "web accessibility means that websites [12], tools and technologics are designed and developed so that people with disabilities can use them". The goal is to ensure equitable access to information and functionality and let them experience the web application and services as a person without a disability. Figure 3.2 speaks for itself and is a beautiful representation of "Equality vs Equity".

The cartoon shows two versions of a scene with people trying to watch a baseball game. The left frame says "Equality", where each person stands on a box, but only the tall and medium-height people can see over the fence. The right frame says "Equity", where the short person has more boxes to stand on, and all can see.

3.2.2.1 Benefits of Digital Accessibility for Smart Cities

1. **Enhanced Inclusivity**: The biggest and most important benefit is that it offers solutions that people with disabilities can utilize. This inclusivity promotes equitable participation and engagement of all residents.
2. **Improved Access to Smart City Services**: Digital accessibility allows people with disabilities to easily access and utilize various smart city services, such as personal banking, restaurant reservations, transportation information, shopping online, healthcare resources, and government services, and anything else that is available on the web. Smart cities that prioritize web

accessibility demonstrate their commitment to inclusivity and equal access for all residents, including those with disabilities. This builds a positive perception and reputation for the city, showcasing its dedication to creating an inclusive and liveable environment for all citizens.

3. **Compliance with Accessibility Regulations**: Almost all countries have accessibility regulations in place that require websites and digital platforms to be accessible. Failing to comply with the standards has repercussions such as bad press and huge lawsuit expenses. Smart cities need to comply with these regulations to ensure equitable access and avoid legal issues. Implementing web accessibility in smart city initiatives helps meet these requirements and ensures that the city is accessible to all residents.

3.2.3 Accessibility Standards and Laws

Accessibility is a fundamental human right and is protected by web accessibility laws because it's a strong expression of political and social awareness. This section covers information about web accessibility laws and their impact on the design and development of digital interfaces.

Web accessibility laws fit into the following three categories:

1. **Civil Rights Laws**: These laws make it illegal to discriminate against people with disabilities under certain defined conditions, such as employment, access to buildings, government services, or "places of public accommodation" such as restaurants, retail, entertainment, etc. Sometimes, these laws include technical standards; other times, they do not. It puts more emphasis on equal rights for people with disabilities.
2. **Procurement Laws**: When making a purchase or when contracting for services. ("Accessibility Laws – DEV Community"), digital accessibility must be considered. For example, the law could state that if there are three potential products and two of them meet accessibility standards, only the products that meet the standards should be considered for purchase. It would be against the law to buy a product that does not meet accessibility standards. The most prominent procurement laws (like Section 508 of the Rehabilitation Act in the United States [14] and EN 301 549 in the European Union [15]) apply only to government entities, but it is possible that a future law could impact private businesses.
3. **Industry-Specific Laws**: Sometimes, an industry is so important to accessibility that the government writes a law just for that industry. Examples include telecommunications and aeroplane travel, both of which have accessibility-related laws in the United States, which are the 21st Century Communications and Video Accessibility Act (CVAA) [16] and the Air Carrier Access Act (ACAA) [17], respectively.

Disclaimer: The explanations offered in this section are for informative use only and do not represent legal advice. Contact your organization's legal representative for official guidance around the legal landscape for accessibility.

3.2.3.1 Civil Right Laws

A civil right refers to an individual's entitlement to freedom, equitable treatment, and protection from discrimination based on specific characteristics. Legislation on civil rights aims to safeguard fundamental rights like privacy and freedom of speech while also preventing discrimination in various areas such as employment, housing, public spaces, and education. To ensure the rights of individuals with disabilities, countries such as the USA and Japan have enacted laws that prohibit disability-based discrimination and safeguard the rights of disabled individuals. People with disabilities are protected with:

1. **The Americans with Disabilities Act (ADA) | United States** [18]: ADA is a comprehensive civil rights law that prohibits discrimination based on disabilities across various domains, including employment, architectural design, transportation, and public accommodation. While the ADA does not specifically mention web accessibility, numerous accessibility lawsuits in the States refer to the ADA's provisions on public accommodation as the legal basis for complaints. This indicates the growing recognition of the importance of web accessibility in ensuring equitable access and opportunities for individuals with disabilities. **The ADA is applicable to**:
   - Private entities encompassing businesses and organizations that engage in leasing, ownership, operation, or rental of public accommodation venues, excluding religious entities and private clubs. In essence, this pertains to establishments open to the public.
   - Both federal and state government organizations.
2. **The Accessibility for Ontarians with Disabilities Act (AODA) | Canada** [19]: AODA is a legislation that covers both private and public sectors in Ontario, including the Legislative Assembly. It applies to individuals or organizations engaged in various activities such as providing goods, services, or facilities, employing individuals in Ontario, offering accommodation, owning, or occupying buildings or premises, or participating in prescribed businesses or activities. The AODA sets out requirements and standards to promote accessibility and remove barriers, ensuring equal access and opportunities for individuals with disabilities throughout the province of Ontario.
3. **The Equality Act of 2010 | United Kingdom** [20]: In the United Kingdom, the Equality Act of 2010 is a significant law that addresses web accessibility. It serves as a comprehensive legislation that prohibits discrimination in both the workplace and society based on various grounds, including disability, race, sex, pregnancy or parenthood status, sexual orientation or identity, and religion or belief. The Equality Act replaced and consolidated several previous laws that focused on specific populations, such as the Sex Discrimination Act 1975 [21], Race Relations Act 1976 [22], and Disability Discrimination Act 1995 [23]. This consolidation simplifies the legal framework and strengthens civil rights protections in the public sector. **The Equality Act applies to**:

- Public authorities and organizations when they are carrying out public functions, provide a mechanism for individuals to file complaints and pursue legal action if necessary.

4. **The Act on the Elimination of Discrimination against Persons with Disabilities | Japan** [24]: One of the actions Japan took to eradicate social barriers for people with disabilities was establishing the Act on the Elimination of Discrimination against Persons with Disabilities in June 2013. The legislation prohibits discrimination based on disability and requires organizations and private entities to provide the necessary accommodations that remove the social barriers people with disabilities face.

3.2.3.2 Procurement Laws

Procurement involves government entities acquiring goods and services from external sources. To ensure responsible spending and prevent fraud and waste, regulations and laws govern government procurement. In the case of technology procurement, specific standards and requirements, including accessibility considerations, are outlined. Two examples of such procurement laws are Section 508 of the Rehabilitation Act and EN 301 549. These laws emphasize accessibility and guide the purchasing process to ensure that technologies meet accessibility standards. By incorporating these requirements, government entities can promote inclusivity and ensure that the technologies they procure are accessible to all individuals.

1. **Section 508 of the Rehabilitation Act | United States** [14]: The US federal government is mandated to prioritize accessibility in its procurement of information technologies, encompassing websites, computers, and all digital interfaces. This requirement extends to both software and hardware, ensuring that accessibility is a fundamental consideration in the acquisition of digital resources. "1194.21 — Software applications and operating systems", "1194.22 — Web-based intranet and internet information and applications", and "1194.31 — Functional performance criteria". Resonates well with Section 508. A noteworthy change that happened in January 2017 was the inclusion of WCAG level A and AA guidelines.
2. **EN 301 549**: **"Accessibility Requirements for ICT Products and Services" | Europe** [15]: The European Union has formulated two directives that require implementation into national legislation by member states. Directive 2016/2102 applies to the public sector, while Directive 2019/882 applies to the private sector, including e-commerce, banking, e-books, and electronics. These directives align with accessibility standards such as EN 301 549, which covers various technologies and references WCAG 2.1 level AA for web accessibility.
3. **European Accessibility Act (EAA)** [25]: The European union formally adopted the EAA on June 2019, which is like the ADA in the United States, documenting accessibility guidelines for EU member states.

3.2.3.3 Technology Laws

It provides legal outlines and legislative regulations for distributing, collecting, and storing digital information. There are numerous laws focusing specifically on different aspects of technology, such as intellectual property, software, hardware, and the Internet. Following are some of the technology-specific laws regarding the Internet:

1. Guidelines for Indian Government Website | India [26]
2. 21st Century Communications and Video Accessibility Act (CVAA) | United States [16]
3. Air Carrier Access Act (ACAA) | United States [17]

3.2.4 Limitations of Existing Solutions

There's not a shred of doubt that the concept of accessibility in smart cities is gaining momentum. However, there are many limitations to the existing solutions. Such limitations shed light on the need to follow accessibility standard Web Content Accessibility Guidelines (WCAG) and be creative. The key limitations of existing accessibility solutions in smart cities are as follows:

1. **Limited Awareness and Training**: The biggest challenge is the lack of awareness and training among stakeholders, including city planners, architects, and service providers. Web Content Accessibility Guidelines (WCAG), set by the W3C Process, ensures that adhering to their success criteria will provide equitable access to the needs of individuals, organizations, and governments globally. Different cities and regions may have varying guidelines and regulations, leading to inconsistencies in implementing accessibility features. Without proper knowledge and standardization, it becomes challenging to design and deliver truly inclusive smart cities.
2. **Affordability**: According to the National Disability Institute [27], On average, households with working-age adults who have disabilities are estimated to require an additional $17,690 per year for a household at the median income level. Assistive technologies such as screen readers, switch devices, and refreshable braille keyboards are expensive and can be a barrier for people with disabilities.
3. **Digital Divide**: Like income divide, digital divide, plays a key role in further exacerbating accessibility challenges. People with disabilities who lack digital knowledge, access to digital devices or weak internet connectivity may face barriers in benefiting from digital accessibility solutions in smart cities.
4. **Limited Spectrum**: Not all disabilities are visible; a person with cognitive disability might appear to be completely healthy physically but experiences challenges understanding complex language and ideas. Some people with cognitive disability easily become annoyed or angry when they sense difficult situations. Too many options can cause them to freeze, disabling them to complete the task. This limited spectrum leaves certain individuals underserved.
5. **Retrofitting Challenges**: The harsh reality is that most of the existing infrastructure was not built with accessibility in mind, leading to difficult and costly retrofitting modifications to buildings, transportation systems, and digital interfaces to meet accessibility standards.

Addressing these limitations requires collaborative efforts from various stakeholders, including government bodies, technology developers, accessibility experts, and people with disabilities. Meriting the limitations and actively working towards overcoming them, smart cities can develop more comprehensive and inclusive solutions that truly enhance accessibility for all residents and visitors.

3.2.5 Web Content Accessibility Guidelines (WCAG)

As per w3.org [6], Web Content Accessibility Guidelines (WCAG) 2.1 [6] provide instructions on enhancing the accessibility of web content for individuals with disabilities. Accessibility encompasses various disabilities such as visual, auditory, physical, speech, cognitive, language, learning, and neurological impairments. While these guidelines address a broad range of issues, it is important to note that they may not fully meet the requirements of individuals with every type, degree, and combination of disability. These guidelines also make web content more usable by older individuals with changing abilities due to ageing and often improve usability for users in general.

3.2.5.1 WCAG Layers of Guidance

W3.org has provided the following layers of guidance to meet the needs of various roles such as policy maker, product owner, designer, developer, and tester:

- **Principle**: Four principles; perceivable, operable, understandable, and robust, at the top, are the four pillars of web accessibility.
- **Guidelines**: Under the principles are the 13 guidelines, which establish the structure or a framework and aim for achieving a greater understanding of success criteria and implementing effective techniques.
- **Success Criteria**: Every guideline is supported by testable success criteria, which can be used to evaluate the solutions. Furthermore, three conformance levels A (lowest), AA, and AAA (highest), are established to cater to the requirements of diverse groups and circumstances. Most successful organizations aim to meet AA conformance levels.
- **Sufficient and Advisory Techniques**: A diverse range of techniques are documented for each guideline and success criteria. These educational techniques can be categorized into two groups: those that are sufficient to meet a success criteria and others that are advisory. The advisory techniques go beyond the required success criteria and are not covered by testable success criteria (Figure 3.3).

3.2.5.2 WCAG Principles

3.2.5.2.1 Perceivable

It means that the web content must be available via any of the three biological senses such as sight, sound, and touch. Website's components, text, and complex widgets must be presented in a way that every user can recognize and understand it.

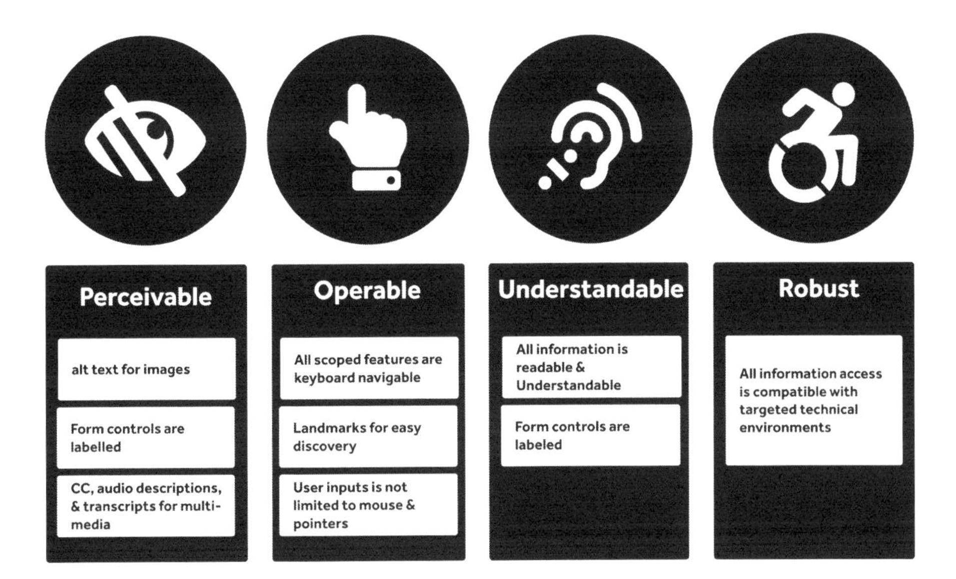

FIGURE 3.3 WCAG principles.

1. **Hearing Web Content**: Most people perceive web pages by looking at them, this works well for people with good vision but not for people with no vision. For them, alternative means such as listening to web content has proven to be a great solution. Screen readers software such as JAWS, NVDA, converts digital text into synthesized speech, allowing blind people to listen to web content, which is an entirely different experience than viewing the content. People who are blind are used to listening rather than seeing, so listening to web content is a logical and natural approach for them.
2. **Feeling Web Content**: Individuals with visual or hearing impairments have an additional sensory modality for accessing web content: touch. Three-dimensional braille characters can be used, either on paper or through modern refreshable braille devices. These devices present text lines one at a time, allowing users to feel and navigate through the content using touch. Tactile output becomes crucial for people who are both deaf and blind, as it is their only means to access web content.

The Importance of Perceivability:

- If a person can't see, all types of information- such as images, colours, backgrounds, visual placement, and layout are completely useless, unless a digital alternative is provided that can be used by screen reader to convert that text either into sound or braille.
- If a person can't hear, audio content is unusable and multimedia content such as videos with sound are less effective, unless a text alternative is provided in either caption or transcript format.

3.2.5.2.2 Operable

It is about making the input methods available to a larger spectrum of people by using input devices, such as mouse or touchpad, keyboard, touchscreen, and voice recognition software.

People with motor skill issues such as tremors in their hands would have a hard time using a mouse or touchpad because such devices require fine muscle controls and precise movements. Furthermore, not everyone can use a keyboard. People with limited movement in their limbs rely on other assistive technologies such as sense movement in their cheek muscles or eye tracking software.

3.2.5.2.3 Understandable

It is about making the content clear and concise and providing functionality that is easy to understand. People with cognitive disabilities like trouble reading complex instructions or remembering details benefit hugely from this principle. Providing consistent placement of top navigation across pages, logical flow of forms and clear labelling, and guidance to complete a task are the characteristics of this principle. On the contrary, if a user fails to comprehend how to navigate a page or what information means, if fails the understandability principle.

3.2.5.2.4 Robust

This principle ensures that a website must be accessible with different combinations of technologies such as using any browser with any combination of a screen reader. Within limits, websites should work across platform, browsers, and devices. Robust principle is the guiding force in a world filled with zillions of gadgets and interfaces.

3.2.6 Role of Web Content Accessibility Guidelines (WCAG) in Building Accessible Interfaces

WCAG provides the technical standards that help meet the accessibility requirements outlined in Section 508. Compliance with Section 508 [14] and adherence to WCAG guidelines go hand in hand because both aim to promote equitable access and usability for all individuals, regardless of disabilities.

In the context of smart cities, it is crucial to ensure that AI-powered interfaces are meeting the minimum requirements set by WCAG because failing to meet the success criteria means that certain sections of society, mainly people with disabilities, are excluded and are denied participation because of the disabling designs and infrastructure. Accessible maps, navigation systems and public transportation apps can facilitate mobility and equitable opportunities to work, live an independent life and contribute to the success of a nation.

3.3 ASSISTIVE TECHNOLOGY

Any piece of equipment or technology that enables a person with a disability to perform functions that might otherwise be difficult or impossible can be called assistive technology [28]. A person with limited hand mobility uses a keyboard to operate a digital interface, or a person who is blind may use software such as JAWS or NVDA that reads text on the screen in a synthesized voice; similarly, people with low vision may use screen zoom software to enlarge the content.

The formal definition of assistive technology was established in the Technology-Related Assistance for Individuals with Disabilities Act of 1988 (The Tech Act) [29]. It was later amended in 1994 and replaced with the Assistive Technology Act of 1998 (AT Act) [30]. The original definition has remained consistent throughout these legislative changes and has also been utilized in the Access Board's Electronic and Information Technology Accessibility Standards, developed under the 1998 amendments to Section 508 of the Rehabilitation Act [31].

People who require assistive technology are:

1. People with disabilities
2. Older people
3. People with a gradual functional decline or mental health condition like autism.

Thanks to the advancement of technology, a diverse range of assistive technology is available today. However, assistive technology works well only when accessible solutions are designed and developed; having the right assistive technology does not guarantee equitable access when solutions are not built with accessibility in mind.

A well-trained professional in the digital accessibility field can help choose the right assistive technology. However, some of the common use cases are as follows (Table 3.3):

A person can have a combination of disabilities, and recommendations for assistive technology vary based on the type of disability. The next section sheds light on how dynamic web applications can be made accessible.

3.4 WEB ACCESSIBILITY INITIATIVE – ACCESSIBLE RICH INTERNET APPLICATIONS (WAI-ARIA)

Modern-day applications are dynamic, where a user gets real-time notifications, chat messages, alerts and dynamic content updated based on the user's action. All of this works well for a person who can visually see the digital interface. However, it's

TABLE 3.3
Assistive Technology Recommendations for Disability Type

Disability Type	Recommended Assistive Technology
A person with no vision but good hearing abilities	Screen reader such JAWS, NVDA, Talkback, or Voiceover
A person with no vision and hearing abilities	Refreshable braille keyboards
A person with low vision	Screen magnifier software such as ZoomText
A person with limited hand mobility	Keyboard or switch devices

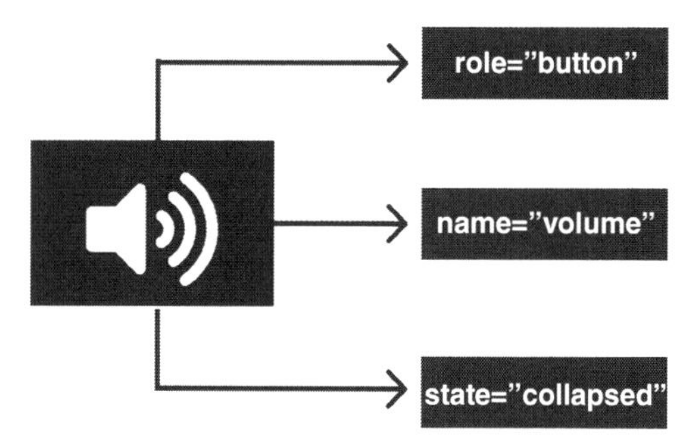

FIGURE 3.4 Role, name, and value of a volume button.

challenging and sometimes impossible for an individual with a vision issue to perceive new information/content updates made to the page. For them, the World Wide Web Consortium (W3C) [7] wrote the technical specifications for WAI-ARIA, commonly known as ARIA, which helps define roles, states, properties, and values for interactive elements to make the content accessible to people with disabilities using a screen reader.

ARIA attributes enhance accessibility by providing additional semantics to HTML elements that can be detected by a screen reader (Figure 3.4):

1. A button's role informs about its intention, for example, "button".
2. A button's name gives it an identity, for example, "volume".
3. A button's state informs about its current setting, for example, "collapsed".

Key feature of ARIA is that it's useful only for Assistive technologies. It does not change anything for sighted users; for example, aria-hidden=" true" does not hide element from visual users, but it does hide it from screen reader users.

ARIA is not a programming language and has no interactivity or procedural logic of its own. Building complex components such as tree menus or a tab panel requires a combination of semantic HTML and JavaScript to achieve content functionality and ARIA to label things properly with names, roles, states, and relationships. ARIA is an API for communicating information to screen readers.

Three Thumb Rules of Using ARIA:

1. NEVER use ARIA unless you must.
2. ALWAYS use ARIA when you must.
3. You're doing it wrong (overusing).

A developer enlightened with ARIA always goes overboard, uses ARIA everywhere, and considers every component a custom ARIA widget. Only use ARIA when there's a compelling reason to build complex and custom widgets that are not available within native HTML elements.

Correct implementation of WAI-ARIA helps screen reader users in the following ways:

1. Landmark regions such as navigation, form, main, and search are rightfully announced.
2. Providing instructions on how to interact with custom components via aria-described by attributes.
3. Announcing the new message or alert banner displayed on the digital interface.
4. Announcing the error messages displayed when submitting an incomplete form.
5. Hiding unnecessary information such as decorative images/icons.

3.5 HUMAN-CENTRED DESIGN

Keeping humans at the centre, solutions to the problems are identified and resolved. Putting focus on the problems faced by people, their challenges are studied, and solutions are designed and developed, sometimes involving them during the design process, ensuring that the solutions effectively meet their needs. A widely adopted and successful AI interface is one that meets the usability needs of all its citizens, which binds them and not divides them, commonly known as the digital divide.

Every smart city must strive to develop a human-centred design infrastructure so that they can develop a deep understanding of their citizens' needs. By putting them at the centre of their decisions, they will develop better solutions. In a nutshell, they must aim to create exceptional, effortless experiences that inspire more citizens to achieve their best in life, health, and wealth.

Key outcomes of a well-implemented human-centred design approach are:

1. Improved user experience for all.
2. Increased usability for people with disabilities.
3. Simplified ease of use, resulting in increased productivity and adoption.

3.5.1 What Is Design Thinking?

Design thinking is the process of creative problem solving, and it recognizes that design can be used to solve complex real-world problems. This methodology prioritizes understanding the needs of individuals, providing a fresh perspective that centres around human experience. It follows a structured framework consisting of five stages, as proposed by Stanford's d. School [32]: empathize, define, ideate, prototype, and test (Figure 3.5).

1. **Empathize**: This stage requires a deep understanding of the challenges faced by people, how do they deal with them, and why do they deal with them this way? It demands to consider a whole person, identify what is influencing them, what matters the most to them, and what needs do they have. It's during this stage that empathy and journey maps are created, interviews are conducted, and questions are asked.
2. **Define**: When problems are identified during the empathize stage, challenges are outlined; this stage clearly outlines and articulates the problem

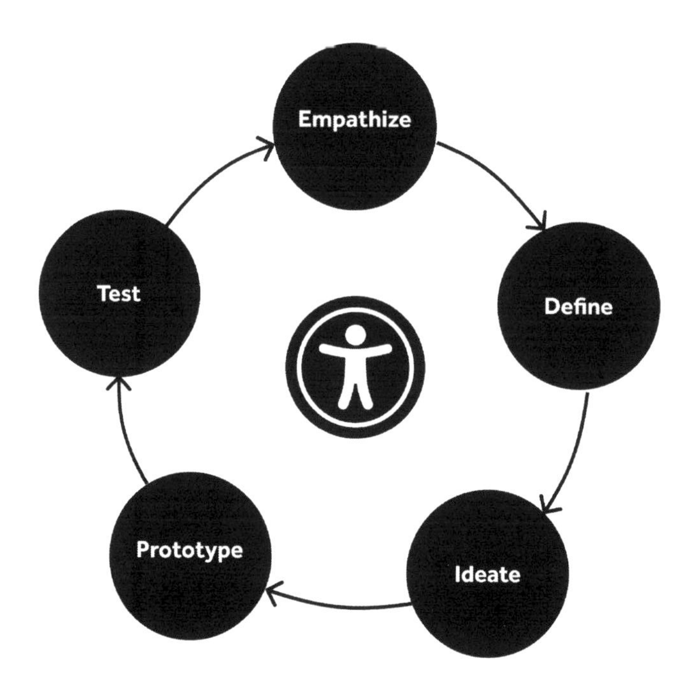

FIGURE 3.5 Five stages of human-centred design approach.

that needs to be solved. Focus is established by framing the problem as a question.

3. **Ideate**: Creative solutions are brainstormed, resulting in a diverse range of options for further development. This stage focuses more on research and inspires innovation.
4. **Prototype**: It's during this stage that the challenges are identified and defined, and solutions are brainstormed, which come to life with a representation of ideas to share with others. Simple user flows for an MVP (minimal viable product) are created for consumers by breaking the larger problem into smaller, testable pieces, leading to market tests.
5. **Test**: During this stage, solutions are tested in short cycles, giving an opportunity to improve and refine the solution. Consumer feedback plays a vital role in making the solution better.

IDEO [33], a renowned global design firm, recognized for its innovation and numerous accolades, introduced the concept of human-centred design. They firmly believe that uncovering the true desires of individuals involves two essential actions:

1. **Observing the User Behaviour**: Understanding people by closely noticing them.
2. **Putting Yourself in the Situation of the Consumer**: Understanding the consumer experience, and to feel what a consumer feels.

The following three lenses can help to evaluate a design process, enabling the success of the solutions:

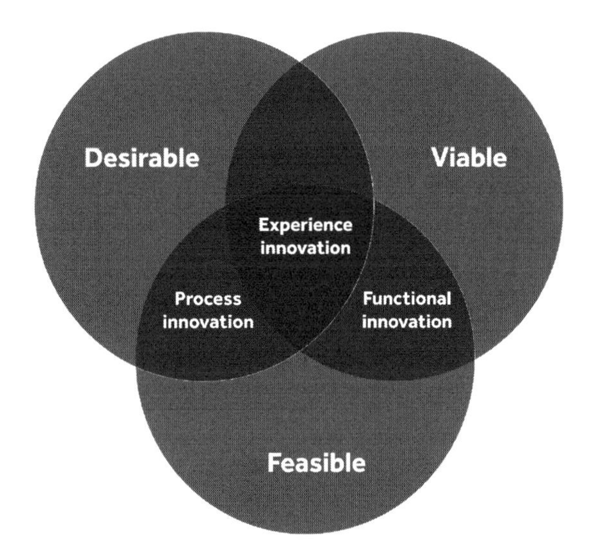

FIGURE 3.6 Three lenses to evaluate a design.

1. **Desirable – The People Lens**: It all begins with desirability. Research, ideate and prototype until satisfactory solutions are not met that meet the consumers' needs.
2. **Feasible – The Technology Lens**: Consider what is needed to make it happen, to bring it to life.
3. **Viable – The Business Lens**: Identify the investment needed and justify its impact on the business (Figure 3.6).

3.6 EVALUATE THE CURRENT STATE OF ACCESSIBILITY FOR EXISTING TRAVEL BOOKING WEBSITES FOR ERROR DENSITY AND TOP ERROR TYPES

The true potential of AI-powered interfaces can be evaluated when the websites built using AI can help a person with a disability. The first few things that a tourist does while planning a trip to another (smart) city is to book a hotel online and figure out the best means(transportation) to reach their destination. It's important to discover the current state of web accessibility for some of the famous travel booking websites to learn how a person with disabilities can access the website and accomplish the task of making reservations.

In February 2019, Web Accessibility in Mind (WebAIM) [34] conducted an accessibility analysis of the most popular top one million domains from the Majestic Million List. Majestic Million [35] is a list of the million most 'important' sites on the web based on the number of backlinks they have.

3.6.1 Methodology

WebAIM's million allows to copy and place a URL into the search area to reveal how websites are using accessibility standards and how it benefits the usability of the site.

- It is important to notice the number of page elements and number of errors detected. This is what gives the error density percentage: anything under 5% is considered very good.
- It is also crucial to focus on the type of errors on the page as this can be used as an opportunity to build capacity in this specific area such as identifying how to respond to these errors and initiating a change in process improvements to address these top errors in the development lifecycle.
- The Wave tool [36], an automated plugin, can detect issues such as images missing alt text, low colour contrast, and missing page structure. This is a great learning tool for discovering what a site is missing and is a good starting point (Figures 3.7–3.9 and Tables 3.4–3.6).

3.6.1.1 MakeMyTrip.com

TABLE 3.4
WebAIM Results for MakeMyTrip.com

Wave accessibility rank	#194,925 of 1,000,000 (top 25% of all home pages)
Popularity rank	#2,054 of 1,000,000
Number of accessibility errors detected	7
WCAG 2 A/AA failure detected	Yes
Number of page elements	339
Error density	2.06% (percent of page elements that have an error)
Top error type detected	Low contrast text and empty link

FIGURE 3.7 WAVE tool results for MakeMyTrip.com homepage.

3.6.1.2 Kayak.com

TABLE 3.5
WebAIM Results for Kayak.com

Wave accessibility rank	#435,623 of 1,000,000
Popularity rank	#2,735 of 1,000,000
Number of accessibility errors detected	35
WCAG 2 A/AA failure detected	Yes
Number of page elements	2,764
Error density	1.27% (percent of page elements that have an error)
Top error type detected	Empty button

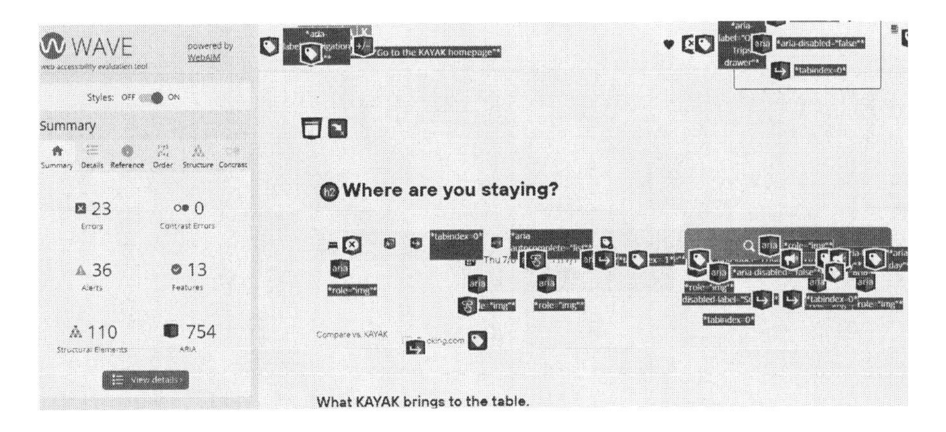

FIGURE 3.8 WAVE tool results for Kayak.com homepage.

3.6.1.3 OyoRooms.com

TABLE 3.6
WebAIM Results for oyoRooms.com

Wave accessibility rank	#237,078 of 1,000,000 (top 25% of all home pages)
Popularity rank	#25,620 of 1,000,000
Number of accessibility errors detected	13
WCAG 2 A/AA failure detected	Yes
Number of page elements	1,347
Error density	0.97% (percent of page elements that have an error)
Top error type detected	Low contrast text; missing alternative text for images; empty link; missing form input label

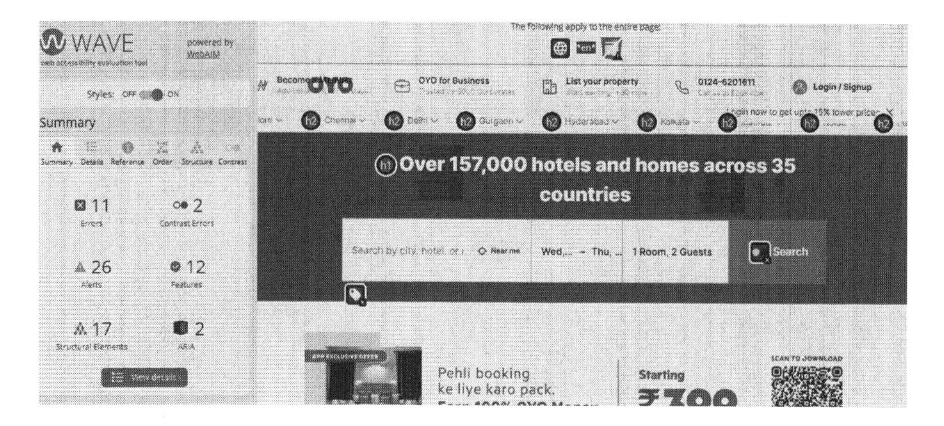

FIGURE 3.9 WAVE tool results for Oyorooms.com homepage.

3.6.2 Recommendations for Design and Development of AI and IoT-Based Interfaces in Smart Cities

The root cause of inaccessible websites/interfaces is not building with accessibility in mind, which leads to retrofitting, and that's both time-consuming and demands high investment. Most successful organizations make the following suggestions part of their DNA, enabling them to broaden their spectrum of consumers who can access their interfaces, resulting in improving the overall user experience:

3.6.2.1 Human-Centric Design Approach

A human-centric design approach is crucial for the success of AI and IoT-based interfaces. Including individuals with diverse disabilities in the design process, conducting usability testing, and gathering feedback at various stages help ensure that the technologies meet their specific needs and preferences. By prioritizing user experience and incorporating customization options, the interfaces can be more inclusive and effectively address accessibility challenges.

3.6.2.2 Bake Accessibility into the Development Life Cycle

It means integrating accessibility at every stage, from planning, designing, developing, and testing. Following Web Content Accessibility Guidelines (WCAG), involving experts, testing using assistive technologies, and seeking continuous improvement, ensures that a digital interface is accessible to all users.

3.6.2.3 Collaboration and Partnerships

Collaboration between government bodies, industry stakeholders, research institutions, and people with disabilities is essential for successful implementation. Public–private partnerships can bring together expertise, resources, and diverse perspectives to drive innovation, address challenges, and scale up initiatives. Collaboration fosters knowledge sharing, fosters a supportive ecosystem and maximizes the impact of AI and IoT-based interfaces.

3.6.2.4 Scalability and Sustainability

Planning for scalability and sustainability is crucial to ensure the long-term impact of AI and IoT-based interfaces. Considering factors such as infrastructure requirements, maintenance, and upgrades, as well as financial models, can help create sustainable solutions. Scalable approaches, such as cloud-based services and modular architectures, enable flexibility and adaptability as the needs and demands evolve over time.

3.6.2.5 Continuous Learning and Improvement

The field of web accessibility, AI and IoT is evolving rapidly, and continuous learning and improvement are necessary to keep pace with advancements. Embracing a culture of learning, conducting research, and staying updated with the latest developments in technology and accessibility best practices are essential. Learning from the successes and failures of past initiatives helps refine strategies, identify areas for improvement, and foster innovation in enhancing accessibility.

By incorporating these recommendations, future implementations of AI and IoT-based interfaces can build upon existing knowledge and experiences. This enables the development of more effective, inclusive, and sustainable solutions that empower people with disabilities, reduce inequalities, and create truly accessible smart cities.

3.7 MYTHS AND MISCONCEPTIONS ABOUT ACCESSIBILITY

Discussing web accessibility makes some people nervous, and they come up with a list of objections, most of which are myths. This section examines and breaks down those myths:

1. **Myth – Accessibility Benefits Only a Small Minority**: The truth is that it benefits a wide spectrum of people; when applications/infrastructures are designed and built with accessibility in mind, a designer/developer thinks specifically about people with disabilities, but its benefit extends far beyond just people with disabilities. For example, maintaining a good colour contrast not only helps a person with low vision but also helps a person operating the device in bright daylight. Similarly, making the form keyboard friendly not only helps a person with limited motor skills but also helps a mom carrying her baby.
2. **Myth – Accessibility is a Short-Term Project**: With the launch of a new AI-powered application every minute, it's safe to say that accessibility is here to stay and requires constant updates. If there are people, there will always be a need for accessible designs. This can be compared to other focus areas such as security or privacy. They will never be phased out and are ongoing requirements in all phases of the development lifecycle.
3. **Myth – Accessibility Should Be the Last Step**: As discussed earlier, designing for accessibility is much easier than retrofitting for accessibility because starting early with accessibility in mind saves both time and money.

4. **Myth – Accessibility Is Hard and Expensive**: When compared to the cost of alternatives (retrofitting, lawsuits, and brand damage), the cost of accessibility sounds reasonable. Building an accessible solution does cost money, but failing to build an accessible solution turns out to be more expensive in the end. Lawsuits are expensive and not to forget about the negative publicity that they bring.
5. **Myth – Accessibility Is Ugly**: When solutions are planned, designed, and built, using the human-centred design approach and industry standards are followed, any solution can be stunningly beautiful.

3.8 CONCLUSION

In conclusion, the integration of AI and IoT with assistive technologies holds immense promise in enhancing accessibility in smart cities, and by leveraging these technologies, we can create a future where everyone has equitable opportunities and participation in urban life. The key findings can be summarized as follows:

- Digital Accessibility is of paramount importance in smart cities ensuring equitable access to the internet services, such as websites, digital interfaces, and kiosks for people with disabilities.
- Human-centric design, usability testing, and customization options are crucial in creating inclusive interfaces and enhancing user experience in AI and IoT-powered interfaces.
- Collaborative initiatives between government bodies, industry stakeholders, and research institutions play a vital role in successful implementations of AI and IoT-based interfaces in smart cities.
- Public-private partnerships and collaborative research initiatives foster innovation, knowledge sharing, and the development of scalable and sustainable solutions.
- Continuous learning, improvement, and staying abreast of technological advancements are crucial in advancing accessibility in smart cities.

REFERENCES

1. "Disability." https://www.who.int/news-room/fact-sheets/detail/disability-and-health (accessed July 12, 2023).
2. A. Houtenville, "Annual Report on People with Disabilities in America: 2023".
3. M. Yin, D. Shaewitz, C. Overton, and D.-M. Smith, "A Hidden Market: The Purchasing Power of Working-Age Adults with Disabilities," 2018.
4. "About ODEP." DOL. https://www.dol.gov/agencies/odep/about (accessed July 13, 2023).
5. "Sustainable Development Goals | United Nations Development Programme." UNDP. https://www.undp.org/sustainable-development-goals (accessed July 12, 2023).
6. "Web Content Accessibility Guidelines (WCAG) 2.1." https://www.w3.org/TR/WCAG21/ (accessed July 12, 2023).
7. W. W. A. Initiative (WAI), "WAI-ARIA Overview." Web Accessibility Initiative (WAI). https://www.w3.org/WAI/standards-guidelines/aria/ (accessed July 21, 2023).

8. S. Triano, "Categorical eligibility for special education: The enshrinement of the medical model in disability policy," *Disabil. Stud. Q.*, vol. 20, no. 4, 2000.
9. Centers for Disease Control and Prevention, "Disability and Health Overview | CDC," September 15, 2020. https://www.cdc.gov/ncbddd/disabilityandhealth/disability.html (accessed July 12, 2023).
10. T. Shakespeare, "The social model of disability," *Disabil. Stud. Read.*, vol. 2, pp. 197–204, 2006.
11. "Microsoft Design." https://www.microsoft.com/design (accessed July 12, 2023).
12. W. W. A. Initiative (WAI), "Introduction to Web Accessibility." Web Accessibility Initiative (WAI). https://www.w3.org/WAI/fundamentals/accessibility-intro/ (accessed July 12, 2023).
13. IISC, "Illustrating Equality vs Equity." Interaction Institute for Social Change, January 14, 2016. https://interactioninstitute.org/illustrating-equality-vs-equity/ (accessed July 12, 2023).
14. "Home | Section508.gov." https://www.section508.gov/ (accessed July 12, 2023).
15. "Web Accessibility Directive—Standards and Harmonisation".
16. "21st Century Communications and Video Accessibility Act (CVAA)." January 27, 2021. https://www.fcc.gov/consumers/guides/21st-century-communications-and-video-accessibility-act-cvaa (accessed July 12, 2023).
17. "Passengers with Disabilities | US Department of Transportation." https://www.transportation.gov/airconsumer/passengers-disabilities (accessed July 12, 2023).
18. "The Americans with Disabilities Act | ADA.gov." https://www.ada.gov/ (accessed July 12, 2023).
19. "Law Document English View." Ontario.ca, July 24, 2014. https://www.ontario.ca/laws/view (accessed July 12, 2023).
20. "Equality Act 2010: Guidance." GOV.UK, June 16, 2015. https://www.gov.uk/guidance/equality-act-2010-guidance (accessed July 12, 2023).
21. "Sex Discrimination Act 1975." https://www.legislation.gov.uk/ukpga/1975/65/enacted (accessed July 21, 2023).
22. "Race Relations Act 1976." https://www.legislation.gov.uk/ukpga/1976/74/enacted (accessed July 21, 2023).
23. E. Participation, "Disability Discrimination Act 1995." https://www.legislation.gov.uk/ukpga/1995/50/contents (accessed July 21, 2023).
24. "The Long Road to Disability Rights in Japan | Nippon.com." https://www.nippon.com/en/currents/d00133/ (accessed July 12, 2023).
25. "EU Web Accessibility Compliance and Legislation." Deque, January 16, 2020. https://www.deque.com/blog/eu-web-accessibility-compliance-and-legislation/ (accessed July 12, 2023).
26. "Introduction | Guidelines for Indian Government Websites (GIGW) | India." https://guidelines.india.gov.in/introduction/ (accessed July 12, 2023).
27. N. Goodman, M. Morris, Z. Morris, and S. McGarity, "The Extra Costs of Living with a Disability in the U.S.—Resetting the Policy Table".
28. "Assistive Technology." https://www.who.int/news-room/fact-sheets/detail/assistive-technology (accessed July 12, 2023).
29. T. D.-I. S. Harkin, "S.2561-100th Congress (1987–1988): Technology-Related Assistance for Individuals with Disabilities Act of 1988." August 19, 1988. https://www.congress.gov/bill/100th-congress/senate-bill/2561 (accessed July 12, 2023).
30. "ATA Fact Sheet 2.15.2021--Fixed.pdf." [Online]. https://rtcil.org/sites/rtcil/files/documents/ATA%20Fact%20Sheet%202.15.2021--Fixed.pdf (accessed July 12, 2023).
31. "Assistive Tech & Communication." https://www.signsofself.org/html/assistive_tech_-_communication.html (accessed July 12, 2023).

32. "509554.pdf." [Online]. https://web.stanford.edu/~mshanks/MichaelShanks/files/509554.pdf (accessed July 12, 2023).
33. "IDEO | Global Design & Innovation Company | Designing Responsibly." https://cantwait.ideo.com/design-responsibly (accessed July 12, 2023).
34. "WebAIM: The WebAIM Million - The 2023 Report on the Accessibility of the Top 1,000,000 Home Pages." https://webaim.org/projects/million/ (accessed July 12, 2023).
35. "Majestic Million." https://majestic.com/reports/majestic-million (accessed July 12, 2023).
36. "WAVE Web Accessibility Evaluation Tools." https://wave.webaim.org/ (accessed July 12, 2023).

4 Analyses of the Present State of IoT and AI Security

S. Karthi, M. Kalaiyarasi, Smita Sharma, M. Vasudevan, S. Saumya, and T. Vedhanayaki

4.1 INTRODUCTION

In the forthcoming generation, supercomputers will lead the universe by outpacing human computing abilities. Due to the advancement in technology, security and privacy measures have been developed, and precautionary actions have also increased for society's growth. Deep learning and machine learning are the most significant algorithms used in Artificial Intelligence systems to promote various sectors like agriculture, health care, transportation, and smart homes. In this, machine learning algorithm is one of the most advanced Artificial Intelligence techniques; it outpaces dynamic grids and does not need an obvious programme design. Machine learning algorithm provides a defensive policy to prevent various attacks and also train the machine to recognize attacks [1]. It spots the initial phase of the attacks and new attacks by using the learning skills and resolves them logically. Internet of Things (IoT) devices receive security procedures from the machine learning algorithm that make them more authentic and approachable. From IoT devices, the data are transferred via fixed sensors and software that are connected to the Internet. This type of device helps humans improve the quality of daily life by making better results and makes the process very easy by reducing human action and upgrading resource consumption. The term "Internet of Things" was introduced in 1999 by Kelvin Ashton to simplify and improve human communication with virtual mechanisms. This technology is used in numerous sectors like smart agrobusiness, health care systems, smart homes, meteorological conditions forecasting, and transportation tracking. The impression of IoT is drastically modifying our connection with technology.

The Internet of Things turns out to be rapidly increasing technology by providing various kinds of skills and amenities with a great impression on societal life and commercial surroundings. Artificial Intelligence (AI) is vitally used in the agricultural sector because it reduces the dependency on human activities. The machine emulates the human action in the field of agriculture. Population will reach to 9 billion, so the requirement of agronomic foods is very challenging, so AI is used to make the potentially effectual and additional lucrative in the agricultural sector [2]. IoT is also adopted by agriculture to increase yield, effectual, worldwide market, and

DOI: 10.1201/9781003459835-4

reduce time and price efficiency. To make fine results and diagnoses for the humanoid body, Artificial Intelligence is used by specialists and other medicinal experts, who use the arithmetic procedures laterally through data science.

When the IoT is developed in the field of health care, it advances the constituent of well-being maintenance. To improve the standard quality of life with unceasing growth with the aid of technology, the smart community or intelligent community is established that steadiness the three effective issues like societal, monetary and ecological advancement using AI and also IoT is used to reduce the humanoid intrusion. In the field of space missions, Artificial Intelligence is used exclusively for Machine learning algorithms, and it discovers several applications such as self-directed vehicles, the well-being of spaceships monitoring and operative managing of satellite assemblages [3]. The important tasks of the IoT and Artificial Intelligence are security and privacy. Some of the tasks and risks that IoT and Artificial Intelligence are inappropriate gadget updates, the absence of effective and strong security etiquette, user obliviousness, and well-known active gadget monitoring.

In order to accomplish data distribution that is linked to the physical units growing with the extensive deployment of sensors in the real universe. For computation tasks, the information is uploaded to the cloud server [4]. The cloud server is used to reduce the storage of the information. Artificial Intelligence systems are planned with no deliberation for safety and are made extremely susceptible to confrontational examples. The confrontational attacks in AI systems can, moreover, arise during the testing and training phase of machine learning. Artificial Intelligence has many security risks including data security risks. AI code reuse and AI supply chain complexity. It is mostly used in cyber security.

Using AI in cyber-security makes faster and more accurate decision-making. AI software, security, and products to identify and detect before they cause damage. Some industries in AI interchange with "Machine Learning". And it also includes the algorithms and the human ability to learn large amounts of information. IoT is the ability to connect to the Internet and exchange information with one another. Device and hardware IoT security and privacy are insecure external ports and network access, lack of security configuration, unauthorized access, etc. Nowadays, IoT technology is interconnected with the devices such as laptops, smart watches, mobiles, and Wi-Fi. Use IoT to protect our mobile phones, cars and home appliances, and use remote to control these technologies. IoT devices communicate with each other and with Artificial Intelligence systems over networks. Network Communication through the agreement like TLS-Transport layer security or VPN-Virtual private networks. It includes strong verification mechanisms such as passwords, two-factor authentication and biometrics to ensure users can interact with these systems.

IoT devices and AI systems should be regularly updated with the latest firmware and security patches to mitigate potential risks. Educating users about privacy and security best practices when IoT devices and interacting with AI systems. Users can make decisions, secure their devices and protect their privacy. Short of safeguarding AI-qualified technologies, numerous societal, monetary and conservational tasks can be cracked through the improved espousal Artificial Intelligence [5].

IoT devices are used to improve security characteristics which are planned based on Fog or Edge computing, distributed ledger technology and machine learning and similarly, some of the open explore complications. Privacy and security are paramount in the context of healthcare and IoT technologies. The integration of IoT devices in healthcare settings offers numerous benefits, such as remote patient monitoring, efficient data collection and improved healthcare delivery. Biometric data and privacy in AI and IoT are areas of significant concern as technologies increasingly rely on capturing and processing sensitive biometric information, such as fingerprinting, facial features, or voice patterns. While biometrics offers convenience and enhanced security, it also raises important privacy considerations.

4.2 ARTIFICIAL INTELLIGENCE IN AGRICULTURE

Agriculture is a serious part of the Indian economy. The average Gross Domestic Product is diminishing gradually from the agronomic sector. Now, it is upsetting that India has converted to an excessive food importer. Smart agriculture is one of the methods to solve this type of situation. The utilization of Artificial Intelligence in agriculture is given in Figure 4.1. Machine learning algorithm makes farming more effectual with highly meticulous algorithms. This type of robots begins with preparation the soil, spreading seeds, weed and pest control, and the process ends with harvesting the crops by identifying their maturity which is known as selective harvesting [6]. To overcome the challenges of the food need of today's increased population, dramatic climatic changes, pandemic situation and insufficient resources availability, precision agriculture or smart agriculture is used with the help of the simulated neural networks, deep net and support vector networks to accomplish the necessities of the agronomic foodstuffs. In many important applications AI robots are used:

FIGURE 4.1 Artificial intelligence in agriculture.

- Soil Preparation
- Yield Monitoring
- Irrigation
- Weather estimation
- Weeding
- Disease and bug management

a. **Soil Management**

Soil management is the significant part in the agriculture, to examine and preserve the soil with accumulation and permeability by spread over the dung and organic fertilizer. In 2018, IBM has constructed a miniscule soil testing systems known as calorimetrical systems that are proficient of finding five indicators. AI's fuzzy logic is used for the exclusion of imprints or ambiguity and unintended hazard studies. It also regulates the superiority of the soil and dirtied soil. Chemical examination is made by the card's micro fluidic chip and humidity level of the soil is identified by the Simulated Neural Networks. Suggestions from the Simulated Neural Network include the movement of carbon model, carbon-di-oxide discharge model, deposited carbon issue sink model, and Bio-sequestration cause model.

b. **Crop Monitoring**

To minimize crop damage Artificial Intelligence made it possible to monitor crop growth. Big data technology is used to select better-yielding and unspoiled seeds. To examine the healthiness of the crop using remote sensing methods such as hyperspectral photography and 3D optical maser scanning for monitoring the crops in the vast area of agriculture. In 1986, the foremost proficient technique, COMAX, was used to inspect the cotton crop, and COTFLEX systems were also used. The fuzzy logic technique is used to survey the soybean crop. MATLAB® algorithm is used to analyze the nitrogen content present in paddy leaves. Computer vision is used to detect the headline time of white and the growth of the chart. Machine learning and threshold segmentation indicate measures of plant growth.

c. **Weeding**

Weeds are known as unwanted plants. For Weed detection, there are many algorithms are used. The basic weed detection process is described in Figure 4.2. The cost of labour increases and the yield of the crops are increased when the weeds are present. The unwanted plant acquires crops in each constraint together with mineral nutrients, water, solar power and space. The machine learning technique is cost-efficient and highly effective.

d. **Disease and Pest Management**

Crop: Paddy

Disease: Fool's Rice, Aspergillosis (Mould)

Machine Vision algorithm is used to identify Fool's Rice or Banana disease. Shift Invariant, long-term memory, Back Propagating Neural Networks are used for the identification of the development of fungoid groups in the paddy crop to prevent from the Mould disease. The basic process involved in paddy best detection is described in Figure 4.3.

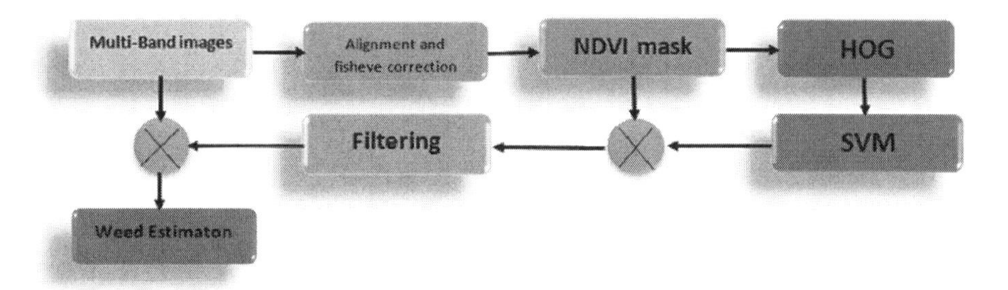

FIGURE 4.2 Weeding.

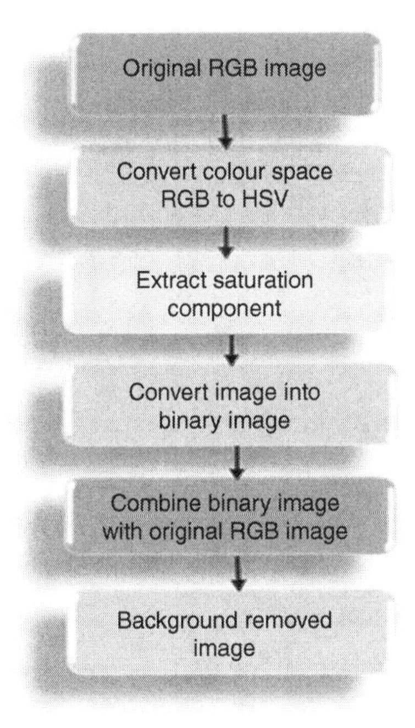

FIGURE 4.3 Paddy pest detection.

Crop: Wheat
Disease: Aphids, SeptoriaLcopersicon, Yellow rust

Support Vector Networks, Canny Edge Detector, Scale and Invariant feature transform, and maximally stable extremal regions are some of the techniques used for the identification of aphid's disease in wheat crops. SVN method is also used for the identification of the leaf spot sternness.

Crop: Soyabean
Disease: Foliar

K-Nearest Neighbour, Support Vector Networks are used for the identification of leaf disease. Raspberry PI algorithm is used for the aerial bug numeration and detection to control the pests that are destroying the crops. Deep convolution neural network is used for pest detection.

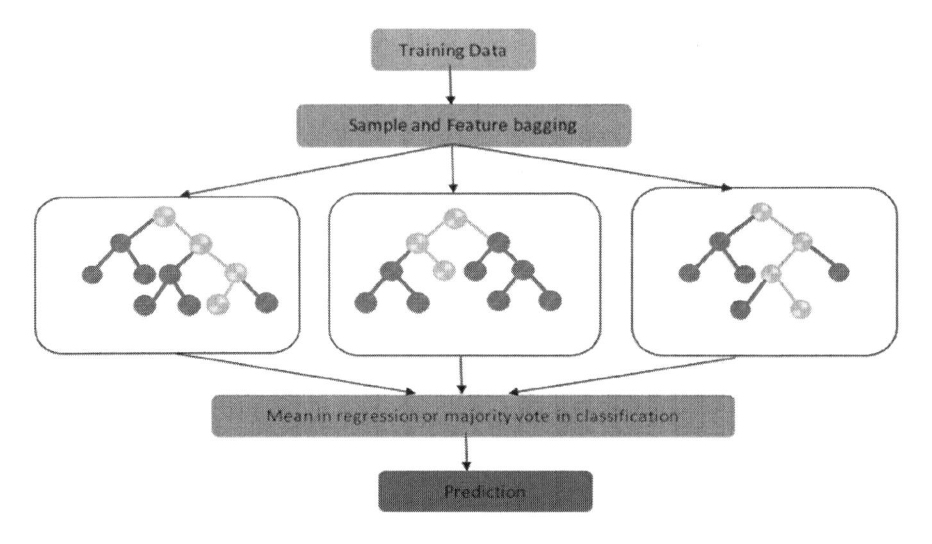

FIGURE 4.4 Feature bagging.

4.2.1 Methodology Used

4.2.1.1 Feature Bagging

Feature Bagging technique is used for the estimation of the high yield at the provincial and international balances with high precision and correctness with the adaptable Predictive Analytics algorithm [7]. A supervised machine learning technique is used for the Feature Bagging with Predictive Analytics algorithm, and it is described in Figure 4.4. To advance the predictive accuracy of the dataset, this method contains the decision trees with the numerous subsets of the given dataset, so the Feature Bagging is the Classifier. This technique contains both the Regression problems and classification.

4.2.1.2 Artificial Neural Network (ANN)

Artificial Neural Network hypotheses the structure of the Humanoid Brain which is based on the Biological Neural Networks using computational problems. Artificial Neural Networks contain neurons like humanoid brain neurons, in which numerous layers are connected to every layer. The basic structure of ANN has been represented in Figure 4.5.

4.2.1.3 K-Nearest Neighbour (KNN)

KNN algorithm is used for the categorization and regression problems. Frequently, the KNN is preferred for categorization problems. K-Nearest Neighbour is one of the modest Predictive Analytics based on Supervised Machine Learning. This algorithm undertakes the similitudes between the new info and existing info. It categorizes the new info into the existing info, which is parallel to them. This K-Nearest Neighbour algorithm is used to categorize the new info into the appropriate group [8]. Figure 4.6 shows how new info is allocated to group 1.

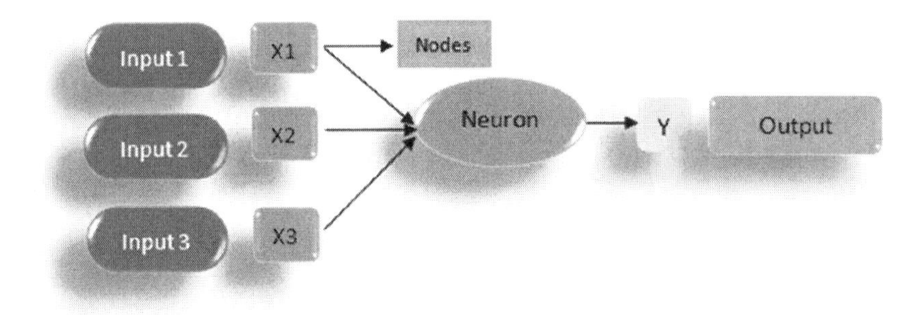

FIGURE 4.5 ANN.

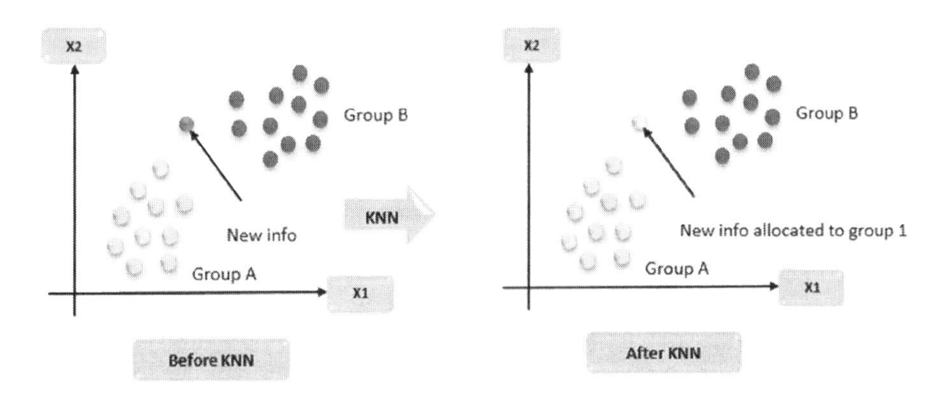

FIGURE 4.6 KNN.

4.2.2 Challenges in AI Agriculture

Even though the advancement of agriculture is improved with Artificial Intelligence technology, still several complications arise. The foremost decisive problem is the non-existence of knowledge and proficiency in Artificial Intelligence technology. To get a result to the problematic, time period is significant to explore, evaluate and investigate. Various tools are required for Artificial Intelligence, like computer hardware, application programmes, detectors, and cameras. Most of the agriculturalists does not have the adequate acquaintance to effectually activate the gadgets due to the lack of training and education in Internet of Things, Machine Learning. The technologists does not regularly effort in the farming grounds, farming practices and explore. The agriculturalists does not have the connection between the Artificial Intelligence technologists, so this one of the complication in AI [9]. One of the most important complication is the cost of the Artificial Intelligence technology equipment. The initial setup of the equipment cost expensive as same and the maintenance of the equipment was also overpriced. Some of the machineries like automatons, unmanned aerial vehicle, cameras, detectors are pricey. So the agriculturalists need to spend much cost of currency to procure and setup the equipments. Sometimes inappropriateoutcomes are also produced by the machinery in Artificial Intelligence. Even though the Artificial Intelligence system has various recompenses, it also creates some additional complications like electronics harsh environmental issues, property exploitation, loss of profession, industry concentration, and ethical perspective. Artificial Intelligence has numerous negatives in the field of agriculture, and several advancements are essential in Artificial Intelligence [10].

4.2.3 Benefits

Artificial Intelligence supports accomplishing the effectual outcomes with the low consumption of efforts to help the agriculturalists hypothetically by leading the agricultural sector effectively. It is possible to compute the volatiles in nearly actual time and also to calculate the outcomes and efforts. AI is effectually useful for carbon-di-oxide gas, aquaculture or soilless culture, and also helpful for animal husbandry. With the help of Artificial Intelligence, a more effective idea is used to simplify the dispersal of the sources. Analysing the precision request according to the food trade minimizes degeneration and expenditure. Artificial Intelligence helps societal and financial agriculturalists like autonomous vehicles, which take food and crops to the market without the cost of the driver. This system also helps agriculturalists to identify the weather forecast in AI apps that recognize the better period for sowing the seed according to the climate condition. Around $43 billion of losses are occurring due to the uncontrolled weeds in the soybean and corn crops, so AI technology can be used to prevent the loseses. To prevent the herbicide resistance in the food crop and to precisely identify the weeds and pests by using the computer vision algorithm. Blue River technology is generating predictive analysis, computer vision and robotic solutions. To minimize 90% the cost of weedkiller for the crop, the precision technology is used to become free of 80% of the chemicals that are actually spurted on the crops. Due to the lack of labourers and the cost that paid for the labourers also too high. In order to reduce these problems, AI robots are used to harvest the crops that are ripened at the right time. It is known as selective harvesting. This method of harvesting done by the robots takes 24 hours to fruitage the 8 acres of agricultural land. Drones are used for crop monitoring that detect bacterial growth and other disease growth on the crop. The images are processed with different algorithms to identify the diseases and provide a detailed report based on the health of the crops.

4.3 IOT IN AGRICULTURE

In various sectors, the Internet of Things has incredible achievements in smart city advancement, health care, space missions, etc., and the field of Agriculture has also become more advanced than previous technologies. Figure 4.7 shows how IoT is used in Agriculture. The Internet of Things does not contain an integrated composition. The configuration is classified into three segments are:

- Object end
- Cloud end
- Network end

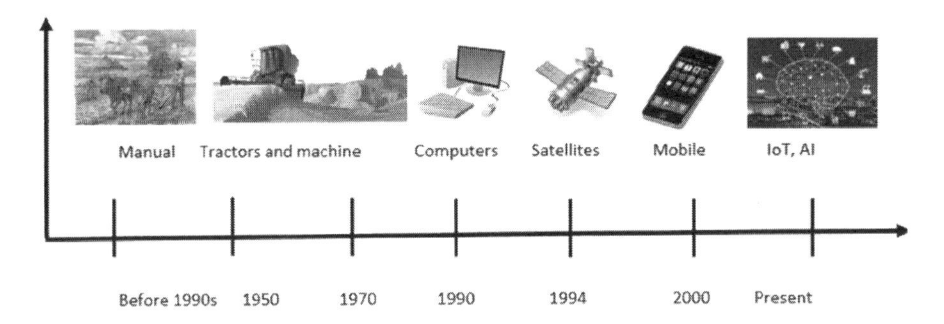

FIGURE 4.7 IoT in agriculture.

To construct the object-end, the physical biosphere is straight to cooperate with the intellectual structure, intelligent gadgets, and intelligent objects. It is also called as front end. The smart scrutiny of detecting information existing in the object end and also establishing the cloud end by performing implementation control. It affords the platform for the synthesis processing sources, memory, and computation. It is also called as back end. Network end is constructed by linking the object end and cloud end with the communication organization.

4.3.1 Methodology Used

4.3.1.1 Wireless Sensor Network

A wireless connection is associatively perception, accumulate, and operation the several data of the recognized entity in the grid range zone from the set of motes which is generally driven by accumulator, and it is several bound rationalization grid organizations. WSNs are classified into two types: terrestrial WSN and wireless underground sensor network. In the agronomic sector, squat depletion exists in the wireless underground sensor network cause of the low rate of recurrence; infrared communication is most commonly used, so the radars are constructed into the farming ground. Terrestrial WSN encompasses smaller receiver dimensions and power utilization than the wireless underground network sensors.

4.3.1.2 Utility Computing

For the advancement of the several Internet of Things sectors, Utility Computing affords the assured platform. Utility computation is raised of distributed systems, equivalent computing, and network computations. For the numerous Internet of Things fields, material structures, accessible stage systems, framework structures, memory structures, and application systems are afforded by the cyberspace-built computation structure. It distributes the users with information storage, a computation that deals with quality of services necessities, memory source pools, and vital observing of implicit computation, as well as reallocation and robust administrations of the system principles. It plays a vital role in the advancement of farming and supports agriculturalists with low-priced stowage information of narrative, representation, audio-visual, and additional farming data that significantly minimizes

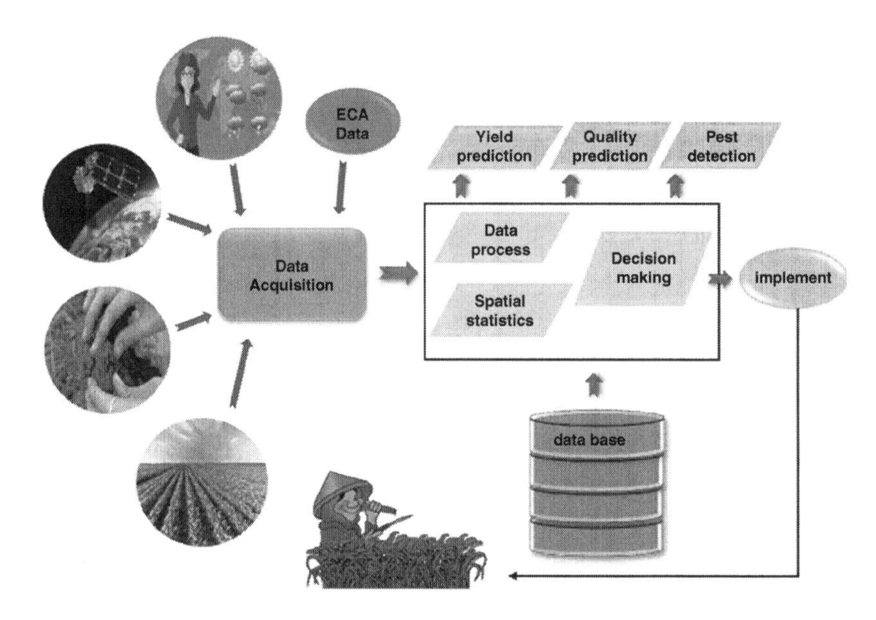

FIGURE 4.8 Big data.

the charge of the farming initiatives. Utility computing assists the feature of extensive intellectual computation, and it is problematic to use primal matters directly in accordance with the methodical level of agriculturalists for decision making. Only agronomic experts can make the exact verdicts and offer recommendations according to the volumetric analysis.

4.3.1.3 Big Data

Big data technology is able to discover the internal relations of composed information, as represented in Figure 4.8. It also modernizes innovative data and affords information benefits a succeeding process. Secured farming generates millions of vital, compound, contiguous information along with farm field directories, carbon-di-oxide atmosphere info, cattle immunization details, and administration venture information. There are some of the frequently used methods to accord through big data technology such as simulation representation, artificial intelligence, imaging visualization, analytical investigation, replication, and topography.

4.3.2 Challenges

In the Internet of Things, there are several challenges in agriculture that should be come across. Some of the challenges are application challenges, grid challenges, security challenges, and the multifaceted carbon dioxide atmosphere, as well as severe problems that should be handled by the numerous gadgets in the discrimination layer. Radars and terminals are severely dented simply by the Heavy sunlight emissions, intense heat, excessive moisture, and heavy fluctuations. Weak recognition and poor control occur due to the regular events of the cattle, which intrude on the duties of the radars and implementation nodes. Consumption of more power and price disturbs the agriculturalists by the attitude of the recurrent substitution of the accumulators due to the data assortment nodes depending on the accumulator through the defined potential to sustain their duties. Due to the uneven and scarce planning of the radars and gadgets, it also handled the diverse situation. Relatively, wireless connections are preferred for farming to the cabled grids to minimize the cost of connecting, and it helps the farmers in numerous ways such as better interacting, being highly pliable, and having good extensibility. In the existing tracking structure, this develops a key software. The wireless connections are also pretentious due to the intense heat, moisture, structural obstacles, and construction due to the various authentic distributions.

The difficulties of surveillance and secrecy are critical complications that occur in farming, such as susceptibilities, grid outbreaks, and secrecy problems. The collective grid assorted congregated network contains difficulties like safety complications, radar grids, portable communiqué grids, and cyberspace, apart from these complications' particular problems like authentication, access control problems, memory, management of data complications, and secrecy defence difficulties of the collective grid assorted congregated networks. Many of the gadgets are interlinked to the substandard assorted network or cloud. This may cause some complications, as well as an information assortment or gadget assortment. The Internet of Things in farming pretentious the extensibility by the gadget assortment. The usage of combination data through representation are retards by the information assortment. For the standardization and deployment in Internet of Things in farming, several hardware and application manufacturers, institutes and related organizations has taken numerous supports, but more development is required for the strong system, etiquette and recognized to link several assorted gadgets and systems. In the field for farming the Internet of Things technology need to be developed in the societal and economic advancement.

4.3.3 Benefits

Agriculture in the Internet of Things refers to the integration of IoT technologies into agricultural practices to improve efficiency, productivity, and sustainability. It involves the use of interconnected devices, sensors, data analytics, and automation to monitor, control, and optimize various agricultural processes. Sensors are placed across farms in IoT-based agriculture to gather data in real-time on variables including plant development, the amount of light, the temperature, humidity, and moisture levels in the soil. Because these types of sensors are network-connected, data may be continuously transmitted for processing to a cloud platform or central server. IoT-powered agriculture transforms conventional farming by utilizing the strength of sensors, linked devices, and data analytics. It enables farmers to collect real-time data, gain valuable insights, automate processes, and remotely monitor farm operations. By leveraging IoT technologies, agriculture becomes more efficient, sustainable, and capable of meeting the challenges of feeding a growing global population.

One of the main uses of IoT in farm goods is automation. With the use of preestablished thresholds and data interpretation, automation may regulate many different aspects of a garden's operations, including fertilizer, irrigation, and insect control.

IoT sensors, for instance, may be installed on irrigation systems to determine how much water crops require by factoring in the quantity of soil moisture that has to be premonetered. This will eliminate the need to water the plant yet lessen overwatering. As a result, with or without on-site visits, IoT-powered agricultural operations are easier to monitor and sustain. These days, farmers may obtain real-time data and receive notifications or even early cautions thanks to web-based technologies. Considering this, folks can keep an eye on crop conditions. By keeping these in mind, individuals may monitor agricultural conditions from anywhere in the globe, spot anomalies, and react to crises and issues more quickly. Because of this, when we consider the Internet of Things, agriculture provides some really beneficial aspects. Increased production potential, more efficient use of resources, reduced environmental risks, and extended crop viability are some of its effects. Better irrigation techniques as well as a decrease in water runoff, phosphorus, and nitrogen runoff, may also be part of their activity. An Internet of Things (IoT)-based pest monitoring system will evaluate the status of the plants in order to identify early pest emergence and, moreover, to reduce the requirement for increasing pesticide application. The second component that may be seen as improving overall efficiency and helping farmers save time is the automation of repetitive tasks like making rounds. "Industrie de l'IoT in agriculture" refers to the use of Internet of Things (IoT) technology in agriculture to increase productivity, production, and resource use. Hence, by monitoring farming activities in real-time and collecting data while utilizing the Internet of Things (IoT), farmers may make well-informed judgements.

4.4 ARTIFICIAL INTELLIGENCE AND IOT IN SMART CITIES

An intelligent community and digital city are used to encourage unfeasible advancement, develop collaboration through the administration, and also enhance the quality of life for the people due to the usage of innovation-based results. The controlling of significant properties, sources, and inner-city movement for actual progressions is achieved by connecting and stabilizing the aspects of societal, economic, and conservational advancement through delegated procedures to define the town as an intelligent community [11]. For the purpose of superior inhabitant communication and administration proficiency, the radar techniques are constructed to assist societal and inter-city interconnectivity using Artificial Intelligence and the Internet of Things with data and interaction knowledge.

a. **Utility Computing**

The initially proposed plan for the Internet of Things structure is utility computing. The information that takes place from the cloud is managed from the numerous mechanisms in the Internet of Things structure. Continuous collective data are retrieved using the distant detecting method by utility computing through a grid. In the absence of hominoid involvement and agenda, data or information could be easily assigned with dynamism. Utility computing contains high expectancy and slow executive methods. Through the various perception gadgets, the assortment of information is operated. It contains the finest proficiencies in relation to information and weak extensibility. It is expensive in grid construction. The source information could be directed to the cloud, so the secrecy is dangerous.

b. **Edge Computing**

In the Internet of Things, much of the data formed through the Internet of Things comes about regarding the perception of termination. Many of the problems could be resolved by edge computing, which is raised from utility computing in the Internet of Things system. Inside the minor section, the assortment information operated. When information is directed by radars and other gadgets, the edge layer could be retorted more rapidly so that the data that is directed would accumulate. This contains many advantages compared to utility computing. It contains highly secured secrecy, is highly vigorous, and has extensibility, in contrast to utility computing. But it is incompetent in comparison to utility computing.

4.4.1 Privacy and Security in Smart Cities

4.4.1.1 Privacy Concerns

Data Collection and Usage: Smart cities collect vast amounts of data from various sources, such as sensors, cameras, and connected devices. Ensuring the privacy of individuals' personal information and preventing unauthorized access or misuse of data is crucial.

Personally Identifiable Information (PII): Smart city systems may inadvertently collect PII, such as location data, biometrics, or health information.

Tagging and Observation: If the public becomes aware of the following techniques for tracking and monitoring technologies and surveillance cameras on a large scale, concerns about abuse and privacy protection for citizens are raised. Keeping in mind how challenging it might be to strike a balance between public safety, security, and the necessary protection of individual rights.

4.4.1.2 Security Risks

Cyber Risks: Hackers can leverage network topologies and Internet of Things (IoT) devices, including smart cities, to their advantage. It should go without saying that these systems are vulnerable to intrusions or outages of vital services, including instances of truly horrific privacy violations and interruptions.

Unlicensed Entry and Management: Since smart city systems are designed for linked networks, they might become the focus of unauthorized entry and management. The possibility is that bad actors would take advantage of these weaknesses to carry out destructive actions that could lead to a network takeover or the interruption of vital services.

Data Accuracy and Credibility: It is critical that smart cities guarantee the data they collect is based on accuracy and dependability. The reliability of the product and network may be harmed by a major judgement failure caused by data life adjustment or fabrication.

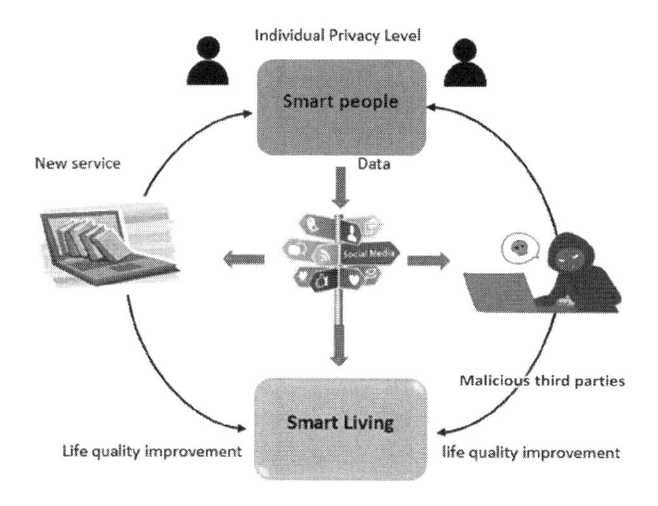

FIGURE 4.9 Safety and security in smart cities.

Accessibility Issues: Preserving the overall safety of the ecosystem requires smooth and reliable interaction amongst these disparate elements. Figure 4.9 shows the link between the level of individual and worldwide privacy.

4.4.1.3 Privacy and Security Challenges

Establishing robust encryption methods and access controls in place to safeguard data while it's in route and at rest, making sure that only authorized parties may access and use it. Integrating privacy concerns into urban infrastructure from the very beginning to reduce the gathering and storage of personal data and guarantee that data minimization guidelines are observed. Getting express consent for data usage, educating the public about the data gathered and processed in smart city systems, and empowering people to make judgements about disclosing personal information. Putting in place thorough cybersecurity safeguards, such as network segmentation, systems for intrusion detection, frequent patching and upgrades, and constant monitoring, to reduce the likelihood of intrusions.

Establishing the transparency, explicability, and lack of biases in AI algorithms utilized in smart city systems in order to uphold public confidence and avoid discriminatory consequences. Fostering cooperation between various parties—government agencies, digital companies, and private citizens—in order to set privacy and security guidelines, exchange best practices, and advance a coordinated strategy to handle issues in smart cities.

4.5 ARTIFICIAL INTELLIGENCE AND IOT IN HEALTHCARE

To take the outstanding verdict in medical management and medications, the doctors are providing various data about the finest therapy strategies and establishing recipient ways with the aid of Artificial Intelligence. The medicine organization and therapy advancement by strengthening in the tedious works to modify intensely initially from a plan of therapy strategies and to several extents in medical management are entrenched with the Artificial Intelligence and Internet of Things [12–15]. Various Internet of Things systems are used in medical management controlled remotely, such as intelligent radars, fitness monitors, abiliment biostatistics radars, glucose level and hypertension detecting medicine chemists, and intelligent cots. This issues the effectual outcomes by subsequent well-being of a recipient by the usage of the recipients' exact abiliment devices. It renovates and directs the individual information of the persons on additional gadgets when the pulse level oscillates or drops beyond personal aid [16,17].

4.6 CONCLUSION

At the cutting edge of development, artificially intelligent IoT technologies, security, and privacy must be taken into account. These are some prerequisites that must be fulfilled or examined before addressing AI and IoT security and privacy. The confidentiality of the data is ensured by the implementation of robust encryption techniques and key management systems that guarantee data integrity and security. Installing robust security measures is necessary to stop the misuse of cyber-physical systems or AI and IoT infrastructures. Make sure you adhere to stringent password guidelines and use secure methods, such as dual-factor authentication, to verify your accounts. Verify that features like IoT-to-IoT and system-to-device connectivity have also been configured correctly. Use secure channels for communication, such as HTTPS, MQTTTLS, or others, and refrain from manipulating or monitoring user behaviour. The phrase "IoT devices" should be changed to "updating the IoT devices' software and firmware" and "patching up security holes and fixing problems occasionally." Put in place safe security procedures that update the software to protect them from the most recent dangers. Before granting access to AI devices, use device authentication services to trace and verify the provenance of IoT devices. To reduce these risks, dependable architecture, stringent test and validation protocols, efficient data security measures, and the availability of organized and understandable AI algorithms are all necessary. The whole life cycle of an AI system has to address security concerns in order to reduce the likelihood of infractions and simultaneously establish a stable environment for the use of AI technology. Smart cities can completely benefit from AI and IoT, but there are still privacy and security concerns that affect individuals and systems. These challenges are anticipated to be resolved in order to fully utilize AI and IoT in smart cities.

REFERENCES

1. Khan, A. A., Laghari, A. A., Li, P., Dootio, M. A., & Karim, S. 2023. The collaborative role of blockchain, artificial intelligence, and industrial Internet of Things in digitalization of small and medium-size enterprises. *Scientific Reports* 13(1): 1656.
2. Ganeshkumar, C., Jena, S. K., Sivakumar, A., & Nambirajan, T. 2023. Artificial intelligence in agricultural value chain: review and future directions. *Journal of Agribusiness in Developing and Emerging Economies* 13(3): 379–398.

3. Vyas, S., Shabaz, M., Pandit, P., Parvathy, L. R., & Ofori, I. 2022. Integration of artificial intelligence and blockchain technology in healthcare and agriculture. *Journal of Food Quality* vol. 2022, Article ID 4228448, pp. 11.
4. Revathi, A., & Poonguzhali, S. 2023. The role of AIoT-based automation systems using UAVs in smart agriculture. In *Revolutionizing Industrial Automation through the Convergence of Artificial Intelligence and the Internet of Things* (pp. 100–117). IGI Global.
5. Elbeltagi, A., Kushwaha, N. L., Srivastava, A., & Zoof, A. T. 2022. Artificial intelligent-based water and soil management. In *Deep Learning for Sustainable Agriculture* (pp. 129–142). Academic Press.
6. Elhoseny, M., Thilakarathne, N. N., Alghamdi, M. I., Mahendran, R. K., Gardezi, A. A., Weerasinghe, H., & Welhenge, A. 2021. Security and privacy issues in medical Internet of Things: Overview, countermeasures, challenges and future directions. *Sustainability* 13(21): 11645.
7. Beniwal, G., & Singhrova, A. 2022. A systematic literature review on IoT gateways. *Journal of King Saud University-Computer and Information Sciences* 34(10): 9541–9563.
8. Javed, A. R., Ahmed, W., Pandya, S., Maddikunta, P. K. R., Alazab, M., & Gadekallu, T. R. 2023. A survey of explainable artificial intelligence for smart cities. *Electronics* 12(4): 1020.
9. Ghazal, T. M., Hasan, M. K., Alzoubi, H. M., Alshurideh, M., Ahmad, M., & Akbar, S. S. 2023. Internet of Things connected wireless sensor networks for smart cities. In *The Effect of Information Technology on Business and Marketing Intelligence Systems* (pp. 1953–1968). Spring Nature.
10. Telo, J. 2023. Smart city security threats and countermeasures in the context of emerging technologies. *International Journal of Intelligent Automation and Computing* 6(1): 31–45.
11. Chong, J. L., Chew, K. W., Peter, A. P., Ting, H. Y., & Show, P. L. 2023. Internet of Things (IoT)-based environmental monitoring and control system for home-based mushroom cultivation. *Biosensors* 13(1): 98.
12. Aski, V. J., Dhaka, V. S., Parashar, A., & Rida, I. 2023. Internet of Things in healthcare: A survey on protocol standards, enabling technologies, WBAN architectures and open issues. *Physical Communication* 102103. vol. 60.
13. Pallathadka, H., Mustafa, M., Sanchez, D. T., Sajja, G. S., Gour, S., & Naved, M. 2023. Impact of machine learning on management, healthcare and agriculture. *Materials Today: Proceedings* 80: 2803–2806.
14. Karthi, S., Narmatha, N., & Kalaiyarasi, M. 2022. Industrial IOT (IIOT) based secure smart manufacturing systems in SME'S. In *Cyber Security Applications for Industry 4*.0. Taylor & Francis. ISBN: 9781003203087. pp. 171–193.
15. Saravanan, S., Kalaiyarasi, M., Karunanithi, K., Karthi, S., Pragaspathy, S., & Kadali, K. S. 2021. IoT based healthcare system for patient monitoring. In *Lecture Notes in Networks and Systems*. Springer. ISSN: 23673389 23673370. pp. 445–453.
16. Devi, V. S. A., Raj, V. H., Kavin, B. P., Gangadevi, E., Balusamy, B., & Gite, S. 2024. An atom quest optimizer for CNN to distinguish IDs in SDN and IoT eco system. In *2024 IEEE International Conference on Computing, Power and Communication Technologies (IC2PCT)*, Greater Noida, India, pp. 1430–1437, doi:10.1109/IC2PCT60090.2024.10486516.

5 Smart City Establishment
AI & IoT'S Role, Vision, and Trends

S. Karthi, M. Kalaiyarasi, K. Deepikakumari, K. Dharani, and Balamurugan Balusamy

5.1 INTRODUCTION

As the world becomes increasingly interconnected and urbanized, the initiation of smart cities has evolved as a transformative solution for sustainable development. There were nearly 511 civic areas with a population of one million people and around 31 big metropolises whose population was less than 10 million people in 2016. To deal with population issue in future, the term smart megacity has been introduced. The term "Smart City" was formulated by IBM. Smart metropolises have emerged as a key action by colourful public sectors in developing metropolises by making them more extensive and sociable to the anticipated community rise and furnishing megacity resides preferable living circumstances. The main purpose of a smart megacity is to optimize and promote further provident and sustainable growth. It's not easy to describe a Smart City, in reality, metropolitan cities are apparently "smart" based on a variety of norms, for example, enforcing-government ventures, improving public literacy gambles and public consultation programmes, focusing on sustainable development and working of Information and Technologies in the near future [1]. Operations that make a megacity smart are smart grid, water operation, waste operation, public care, safe megacity, and smart business operation. Technology plays a vital part in smart metropolises.

Since AI-grounded smart metropolises have been growing fleetly in recent days, AI-grounded operations provide robotization and effectiveness and also solve sequestration, legal, and ethical issues; most smart megacity schemes and technologies are on the basis of creating data and acquiring new information about a megacity's complexity and dynamics. To achieve the climate pretensions and accelerate the energy transition, mass addition and construction workshops will be ineluctable. AI allows original governmental bodies, construction establishments, mileage associations and others to deal with these challenges. Smart metropolises also use IoT bias similar to linked detectors and measures to assemble and access the data. After collecting the data, it's used to ameliorate the services, structure and public serviceability of the megacity. AI takes metropolises to the next position by allowing them to use that data and knowledge for proper decision-making. IoT forms the specialized backbone of every smart megacity in the world.

DOI: 10.1201/9781003459835-5

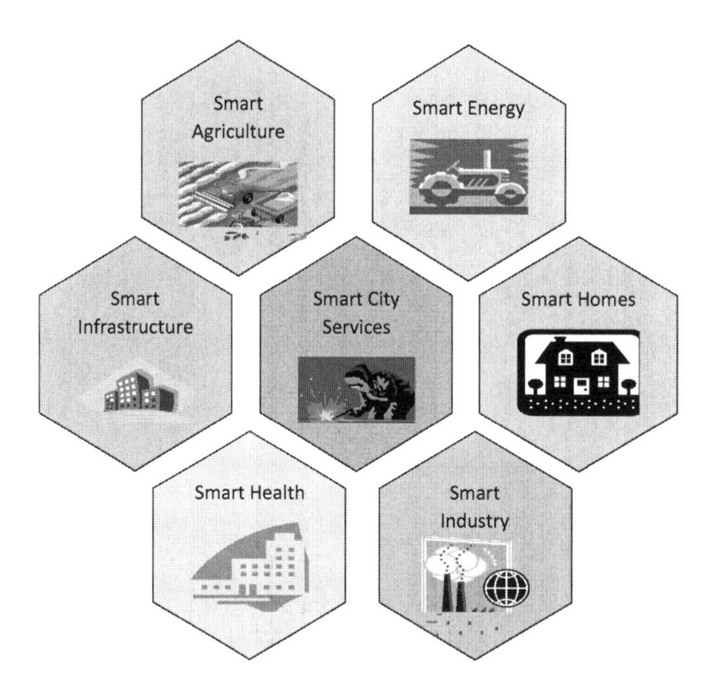

FIGURE 5.1 Outline of the smart city.

Smart metropolises use the Internet of effect to achieve sustainable provident development and improvements. Consequently, IoT systems play a major part in the deployment of large-scale miscellaneous architectures. There are numerous IoT technologies used in smart metropolises such as radio-frequency identification (RFID), near-field communication (NFC), low-power wide-area (LPWA), wireless sensor network (WSN), and DASH7. Like AI and IoT, there are numerous transformative technologies that impact society. The combination of high-speed-quiescence connectivity and technologies similar to the IoT and Artificial Intelligence will enable the metamorphosis towards sustainable smart metropolises. In this chapter, we will examine the vision and trends of AI and IoT in smart megacity establishment. Figure 5.1 provides the figure of the smart megacity.

5.2 ROLE OF AI & IOT IN SMART CITY

Smart meters, which measure and analyse energy usage using IoT sensors, are a perfect illustration of how AI is transforming cities into smarter ones. Smart meters provide city managers access to real-time data to optimize energy use and save expenses.

5.2.1 Smart Grid

The Internet of Things smart grid allows for two-way communication between linked devices that can detect and respond to stoner needs. The goal of the smart grid is to accommodate and control the transportation of power from all sources of generation

TABLE 5.1
Difference between Traditional Grid and Smart Grid

Parameters	Traditional Grid	Smart Grid
Communication	One-way and local two-way communication	Global communication
Monitoring	Blind	Self-monitoring
Operation and maintenance	Check equipment manually	Monitor equipment remotely
Metering	Electrochemical	Digital/microprocessor
Generation	Centralized	Centralized and distributed generation

in order to satisfy the various electric needs of end users. The smartest method to solve this problem is for governments to replace conventional power systems with smart grids. With the increasing demand for electricity throughout the world, we need to discover smarter solutions for producing, storing, and distributing electricity at the right time. The distinction between a regular grid and a smart grid is illustrated in Table 5.1.

IoT digital detectors (power meters, voltage detectors, and fault sensors) keep an eye on the transmission networks. Smart grid is primarily made possible by the Internet of Things, since many of its technology and infrastructure components are IoT based. To optimize power output and maximize the usage of renewable energy, combine environmental data and IoT technology. Through demand-based pricing, Artificial Intelligence optimizes the transmission of power and encourages conservation. The flinging impact of electrical network failures is also predicted by Artificial Intelligence.

5.2.1.1 IoT-Based Smart Grid Technology

Charging stations and smart storehouses are extensively used by everyone currently. It's substantially used for tone-containing renewable energy systems for home use. Detector-enabled IoT bias, appliances and capitals that govern a smart home or any other connected terrain give information on energy consumption. This information is also utilized to cover energy use, cypher costs, ever manage appliances, make cargo distribution opinions, and discover faults. Detector technology plays a critical part because it's these factors that allow consumers to track their energy consumption. Detectors in smart operation allow monitoring and control of homes. Smart measures prisoner data on energy consumption and displays a complete picture of energy operation in the home, including loads and estimates.

5.2.1.2 AI-Based Smart Grid Technology

AI is the fastest growing branch of the high-tech assiduity. The power grids now collect energy from different sources, and the decentralization of the grids and operating the grids has become more complex. This requires massive data. Artificial Intelligence helps process this data as snappily and efficiently as possible, thereby bringing stability and effectiveness. Smart measures process information that can be related to a person and be sequestration sensitive. In electricity trading, AI helps improve forecasts. Machine literacy and Neural Networks play an important part in perfecting forecasts in energy assiduity [2]. Failure operation without regular checks on power outfits and outfit failures are common. Using AI to observe outfits and describe failures before they are can save time and lives.

5.2.2 Smart Traffic Management

With the swift rise of population and urban development in metropolitan cities, traffic blockage frequently happens on roads. To handle these issues, to deal with traffic on roads and to assist the authorities in convenient planning, a traffic management system using IoT in smart cities is highly achievable. It has been identified that, to date, the current traffic management systems may crash. In addition, there is less focus on fluctuations in traffic flow. Therefore, the proposed system manages the traffic on local and centralized servers by exploiting the concepts of IoT and Artificial Intelligence together [3]. The goal of smart traffic management is to make life better on the road. IoT and AI play an important role in smart cities in dealing with traffic. Traffic is detected using sensors, where inter-vehicle and vehicle-to-infrastructure communication is performed using wireless communication devices. Information from the pedestrians and vehicles and requests are processed by the control panel using the First Come, First Serve method.

5.2.2.1 IoT-Based Smart Traffic Management

Smart business signals may look like a typical stoplight, yet they use an array of detectors to cover real-time business. Generally, the thing is to help buses reduce the quantum of time spent idle. IoT technology enables colourful signals to communicate with each other. This is while conforming to changing business conditions in real time. The outgrowth is lower time spent in business logjams and indeed reduced carbon emigrations. The detectors are used to capture business movement, road light detectors, detectors at-risk cells, and detectors installed in public transport systems. Smart measures and mobile apps make on-road parking spaces fluently accessible with instant announcements. Motorists admit caution whenever a parking spot is available to reserve it incontinently [4]. A business monitoring system using IoT technology enables exigency askers to speed up the care medium in case of accidents late at night or in insulated locales. The detectors on the road describe any accident, and the problem is incontinently reported to the business operation system. With every vehicle acting as an IoT detector, a devoted app can make suggestions, determine optimal routes and give advance notice of accidents or business. Further, it can indeed suggest a stylish time to leave. It's each because of a robust algorithm that helps reduce driving time with intelligent business lights.

5.2.2.2 AI-Based Smart Traffic Management

Artificial Intelligence is fleetly changing the world around us, and one of the most effective results it provides is smart megacity technology similar to Parking operation and Business control system. AI is used by numerous nations to make business lights smarter and more effective for securing passengers and on-road motorists. It uses computer vision to describe the viscosity of vehicles on the road and passengers near the path and gives motorists suggestions grounded on the collected data. Videotape Analytics technology is used to describe events and give sophisticated cautions for any unwanted business-related incidents. This technology makes use of CCTV cameras and AI to understand the monitoring zone and to help prevent accidents between two incoming vehicles by estimating the distance between them. Automated vehicles are becoming common these days; while they're frequently allowed as tone-driving vehicles, they're much further than that. Videotape analytics also separate people with and without helmets. Law enforcement officers can work this technology to catch business imminence and over dalliers. It can also be used to control transportation systems in metropolises and public roadways.

5.2.3 Water Management

Water is at the centre of sustainable development and is vital to socioprofitable growth, healthy ecosystems, energy and food products, and the very survival of mortal beings. Water operation and other water-related activities are minor in the scale of public understanding and government involvement of various countries. A few of the major advantages of the operation of water in smart cities are water services, understanding of the water system, leak identification, protection, and surveillance of water quality. A smart water management system tends to further support water utility, diminish financial losses, and validate revolutionary profit models to assist the civic and pastoral folk more. Non-profit water is water that has been produced and is "lost" before it reaches the client. Figure 5.2 shows more effective metropolises in terms of non-revenue water in the 1990s.

5.2.3.1 IoT-Based Water Management

A water management system that incorporates IoT and data analytics is intended to reduce water waste while simultaneously improving water availability. The data for the study is received from the city's water delivery system, as well as the city's water reservoirs and tanks, and is gathered continually by utilizing sensors and transferred to the storage. The head of the water tank has an RF-05 sensor, a distance sensor that

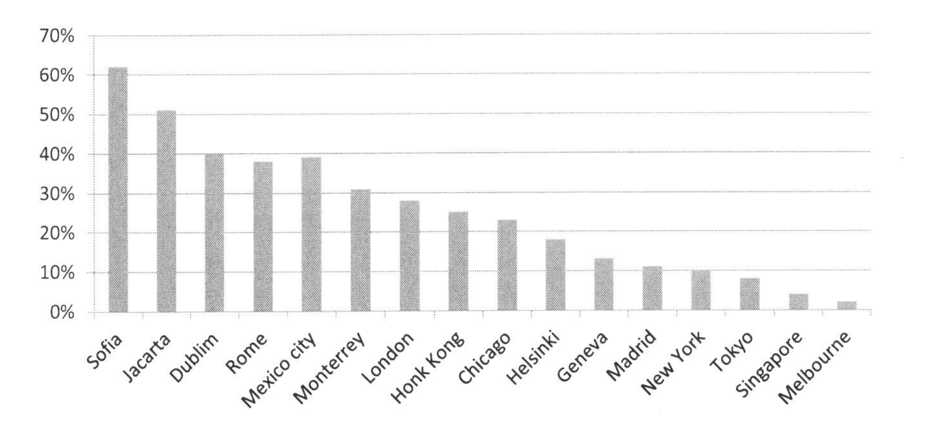

FIGURE 5.2 Non-revenue water more effective metropolitan cities in the 1990s.

measures the volume of water there using SONAR technology. Some hall effects sensors are used to determine the accurate flow of water from an economic standpoint [5]. The chip-based rain sensor finds out whether or not it is raining. It is essentially linked to the Arduino board in order to find out the presence of moisture level on the main board.

5.2.3.2 AI-Based Water Management

With the use of data science, regression models, and algorithms, AI may greatly simplify the process of water management systems. These cutting-edge technologies may be used to construct efficient water systems. AI predicts water use and is used to optimize rainwater gathering. AI prevents disturbances and enables dispersed sensor networks for flood planning. AI may be used to minimize contaminants in water, reducing water pollution and the shortage of clean water. Because AI relies on optics, it may be used to identify the number and composition of harmful substances, increasing the efficiency of waste management systems. Quality of water may be continually surveyed, and instantaneous data on quality can be obtained using machine learning and big data [6].

5.2.4 Waste Management

Trash collection optimization and monitoring have gotten increasingly complex. You can simply track your garbage containers with the aid of smart sensors embedded in smart waste bins. Smart sensors enabled by IoT technology will give you real-time data about your assets. Activities of waste management are not restricted to garbage from households, business facilities, and so on. Waste is a severe challenge that affects the health and hygiene of all city residents. Wastage and sanitation are both intimately tied to cleanliness. So it is not only Smart cities that have become the validates for Solid Waste Management but also many other schemes like Swachh Bharat Mission have also played an important and crucial role in providing public awareness on waste management. The problems and issues faced by current urban waste management can be dealt with through mechanisms such as IoT, AI, and intelligent transportation systems. Current urban waste management challenges and difficulties may be addressed with systems such as the IoT, AI, and intelligent transportation systems.

5.2.4.1 IoT-Based Waste Management

The sensors are used to monitor rubbish pickup and to detect things at disposal sites. Dumpster rental technology is a good example of an IoT-based trash management system. It, like smart bin sensors, may employ a sensor system to track and schedule unique, usually on-demand, waste collection activities. RFID tags, sensors, actuators, wireless sensor networks, near-field communications, and GPS are used in waste management to annotate and share information. IoT advances have improved the current waste management system. Farmers may remotely monitor and control numerous parts of their crops and irrigation systems with IoT sensors in real time. Ultrasonic and proximity-based sensors can alert collection trucks to full bins, assess segregation levels, and determine whether a driver should stop or continue. Weight sensors can also be used to track fill levels in larger bins, which can help farmers

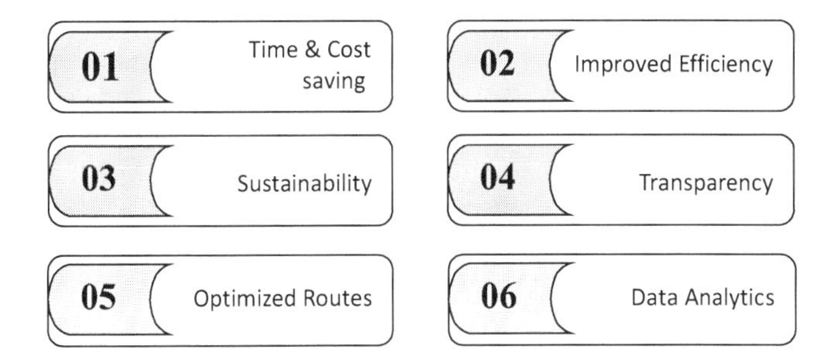

FIGURE 5.3 Reasons why cities need smart bins for waste management.

better manage their use of water and fertilizer, prevent overwatering and runoff, detect pests and diseases, and increase crop yield and quality. Figure 5.3 shows reasons why cities need smart bins for waste management [7].

5.2.4.2 AI-Based Waste Management

Artificial intelligence can separate recyclable materials from non-recyclable ones and monitor waste levels at public garbage cans to improve trash collection. AI is revolutionizing how we transport, collect, and sort all forms of trash, including biohazardous waste and medical waste. As waste management processes become more complicated, robotics, AI, and machine learning have improved worker health conditions and process quality. When waste is being disposed of, a smart recycling container keeps track of and separates it. To effectively and precisely sort garbage, it makes use of artificial intelligence, computer vision, robots, and machine learning. Garbage is disposed of into bins that are scanned by cameras and have the data relayed to servers.

5.3 VISION OF AI AND IOT IN SMART CITY

5.3.1 Vision of AI in Smart City

The aim of "Smart Cities" is to design the metropolitan area of the future that is secure, safe, ecologically friendly, and efficient thanks to all structures, including those for transportation, power, and water. The employment of AI in smart cities has sparked innovation in public transit. This technology enhances punctuality and customer happiness by enabling users of public transit to access and receive real-time updates. Additionally, automated buses are intended for usage in urban areas where they can enhance routes, boost frequency, and cut emissions. AI and machine learning enable governments and cities to make informed decisions that are best for the environment when it comes to pollution management and energy use. AI is also used in smart cities to detect CO_2, which can subsequently influence transportation choices. Smart cities powered by AI are expected to have a significant impact on how inclusive and sustainable urbanization may be. Smart cities powered by AI are establishing new standards for architecture, economy, infrastructure, and more. Cities can conserve energy by using lamps only when a road is in use or include renewable energy sources like solar power, thanks to sensors built into buildings and infrastructure networks.

Better data and analytical tools can influence decision-making and improve urban management: sensors, smart cards, and digital cameras input real-time data into integrated management systems. Crime analytics greatly benefits from AI. The Internet of Things enables all perceptible external items to be automated and intelligent so they can gather and distribute information as needed. The idea of "Smart Homes" has already been a reality for a sizeable amount of time. For instance, even if you forgot to lock the front door when you leave the house or the gas valve wasn't shut off, you may still operate your home using a smartphone. Your home's appliances, such as the refrigerator, air conditioner, microwave, and electric lighting, can be monitored and controlled [8]. Even better, you can let the AI take over and manage them while you are away. Smart street lights that are managed by a central power grid can change their brightness based on how many people are present in the area at any given time. By turning off lights where they are not needed, this aids in energy conservation for the city administration. Real-time data on vehicle density can be gathered by AI-powered traffic cameras and provided to the traffic control centre to help modify the timing of the signals.

5.3.2 Vision of IoT in Smart City

Smart cities are developed by the vast utilization of the Internet of Things. IoT combined with software platforms, Graphical User Interface (GUI), and communication networking, helps in enabling the functions and effectiveness of smart cities. Radio-frequency identification (RFID) devices uses the frequency of radio to transfer the data and to track and people. RFID also helps in identifying the objects. The Internet of Things and Smart City are also working together to reduce environmental problems like poor quality of air and noise pollution and building automation. There are different types of sensors used for monitoring noise. They are microphones, accelerometers, and geophones. Figure 5.4 shows the vision of IoT in the smart city [9].

5.4 TRENDS OF AI IN SMART CITY

AI trends in smart cities are Smart Energy Metering, Park Benches which are IoT enabled, Augmented Workforce Knowledge, chatbots with improved commercial quality, Security Cameras with intelligence knowledge, Improvement in Traffic Control etc,

5.4.1 Smart Energy Metering

One of the applications of AI and ML is smart meters, which have high potential in utilizing filed and energy. Even though smart meters are utilized in small-scale industries, by utilizing them, the consumer can modify their energy requirements and also help in reducing costs. The data generated using AI can help private organizations make a very precise decision about future requirements and present needs.

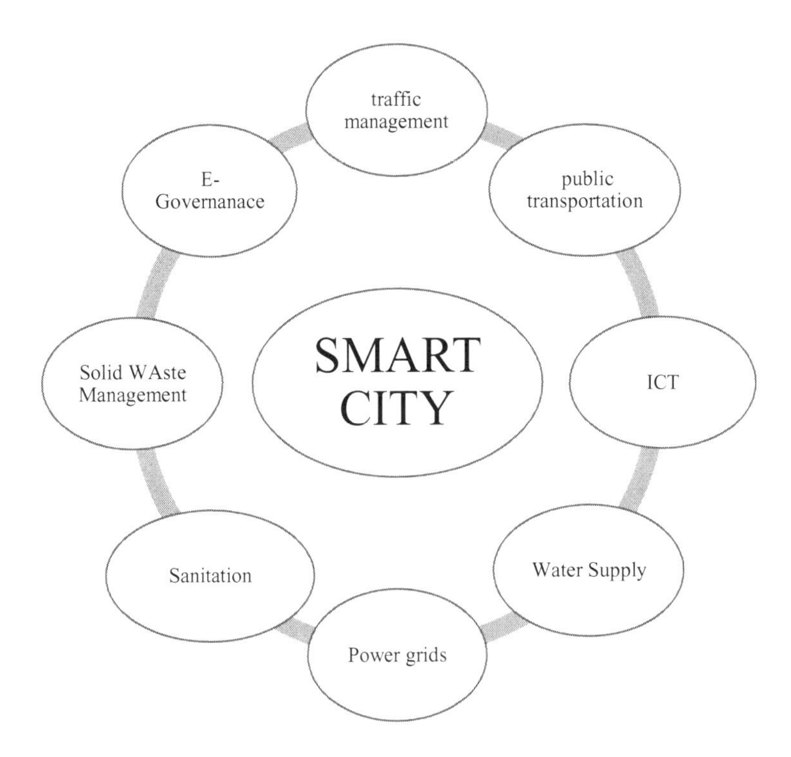

FIGURE 5.4 IoT vision in smart city establishment.

5.4.2 IoT-Enabled Park Benches

The new park benches which are made innovative using IoT are developed to collect a wide range of information that are happening in the environment, and it is analysed with the help of AI. They give intense information about the air quality in the environment, atmospheric pressure and temperature. By collecting these data, appropriate measures are taken to save the present and future environment. By using some apps, we can get the feedbacks of the facilities in the parks. With the use of feedbacks, the government can take remedy measures and improve the facilities in the parks. In Paris, the IoT-enabled benches provide a Bluetooth connection spot.

5.4.3 Augmenting Workforce Knowledge and Skills

The artificial intelligence machine gives suggestions to the technicians and the machine handlers about how to handle and operate the machines. The feedback is also collected from the machine handlers to rectify the errors and mistakes that are encountered in the machines. This helps in providing a better workforce and making better future requirements. The intelligence machine also helps in providing a better way to perform the critical tasks

5.4.4 Intelligent Security Cameras

AI-enabled security cameras are widely used in private organizations and business sector. For example, if we want to detect a person who is wearing "black shirt and white hat", AI can easily detect and sent the alarm or message to the owner [10].

5.4.5 Improving Traffic Control

By combining the concept of Artificial Intelligence and Transport systems, many accidents and speed drivers are captured and warned. Fines are generated by AI for the people who are disobeying the traffic rules and violating it. By this concept, accidents are reduced by a vast amount. AI-enabled traffic lights automatically give signals, which helps in reducing the effort of manpower.

5.4.6 Cloud AI-Powered Robots

The first humanoid robot developed by cloud AI is the XR-1 humanoid robot. In the current version, it performs human actions like opening and closing the door, shaking hands and folding clothes [11]. The social humanoid AI-enabled robot named "Sophia" makes human expressions and interacts with people. Figure 5.5 indicates smart cities in the era of IoT and AI.

5.5 TRENDS OF IOT IN SMART CITY

Edge computing, 5G connection, AI integration, blockchain security, industry-specific solutions, sustainability, improved data analytics, and increased interoperability are all potential developments for the IoT in the future.

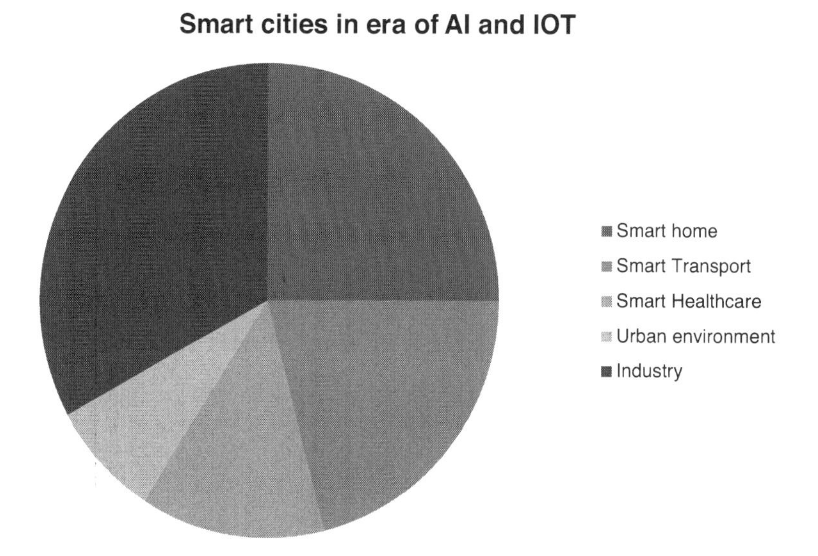

FIGURE 5.5 Smart cities in era of IoT and AI.

5.5.1 IoT in Healthcare

The Internet of Things devices collect and transmit health data, including blood pressure, glucose and oxygen levels, weight, and electrocardiograms (ECGs), to the cloud. This data can then be sent to a designated party, such as a doctor or insurance company, who can access it from any location or device.

5.5.2 IoT in Energy Saving

Systems for real-time monitoring are being developed to assist in keeping track of energy use. As a result, the analyses are rapid and simple, and the findings appear quickly. Early warnings of issues with temperature and vibration in electric motors are provided by an IoT-based monitoring system. Because the motor is prevented from reaching its threshold limit by IoT-based conditional monitoring, it lasts longer and has fewer failures [12].

5.5.3 IoT in Traffic Control

The wireless sensors, RFID tags, and BLE beacons deployed at the traffic lights as part of this intelligent system allow it to track the passage of cars. The digital roadmap, which is GIS-enabled, is linked to control rooms through a real-time data analytics tool for traffic monitoring.

Utilizing a digital image processing approach, the intelligent traffic management system takes pictures of the moving cars at the traffic lights. Wireless sensors then transmit this data to the control panel. The system uses RFID tags or BLE beacons to track vehicle movement, reduce traffic congestion, find stolen vehicles, and even clear the way for emergency vehicles equipped with RFID readers.

5.5.4 IoT in Supporting Security

Traffic events are discovered using Automatic Incident Detection. These systems are employed on roads as well as in tunnels for safety reasons. In addition, Internet of Things technology dissects data to give insightful knowledge that helps communities make decisions regarding public safety.

5.6 CHALLENGES OF AI AND IOT IN SMART CITIES

The digitalization of every element of our life is what the Internet of Things promises. Considerable problems must be taken into account while developing IoT systems for smart cities and deploying them.

5.6.1 Lack of Infrastructure

Modern, highly modern infrastructure is required for smart cities, and each part of the machinery must be linked to the Internet for monitoring. Better judgements for increasing urban assistance for residents, linked IoT devices fetch data for IoT from the physical environment. Due to population expansion, there is an increased need for housing, healthcare facilities, educational institutions, and entertainment venues. All the residents of smart cities who will use the IoT infrastructure should have their infrastructure demands met by AI and the IoT. Sustainable development that is closely tied to framework development in rising countries has significant implications for artificial intelligence. Additionally, the old (existing) infrastructure should be connected and used for a variety of purposes by AI and the IoT. IoT technology have made a substantial contribution to the majority of the specific components and infrastructures of smart cities. There are several economic prospects and significant development potential since smart city infrastructure and IoT technology share many fundamental principles and ideas.

5.6.2 Inadequate Funds or Capital

A research discovered that money and forecast were crucial before starting any project. To create, advance, and sustain smart city development, the public and private organizations need provide adequate funding. Lack of funding leads projects to take longer to complete, which raises project expenses once again. According to Abdalla, the major risk to the development of smart city plans is a lack of capital finance and investment. Projects should be prioritized using AI and IoT approaches based on their importance, needs, timeframes, and other factors, and money should be distributed as efficiently as possible using the best optimization methods. The government may require extra money from businesses in the commercial sector that are enthusiastic about these smart city initiatives in order to properly solve this issue.

The interconnected networks that make up smart cities produce vast volumes of data rich in knowledge, foster innovation, and link businesses, governments, and individuals. The information lays the groundwork for running cities in a way that will increase their efficiency and sustainability. Large-scale data sharing and storage pose a number of issues and difficulties. Private information about residents, official records, and data from all private organizations might all be included in this. However, cyber security issues lead to worries about data security and privacy dangers to smart city network. For instance, a leading US pipeline network operator suffered a cyber-attack, forcing it to temporarily halt operations. To restore operations, the company paid the hackers a sizeable sum of money. China enacted a new legislation governing the protection of personal data on August 20, 2021, and it will go into effect on November 1, 2021. According to this legislation, people's separate consent must be sought before sensitive personal information including biometrics, medical histories, bank accounts, and locations are processed. Additionally, it forbids the collecting of private information, such as biological and facial details, from people when writing in specific limited circumstances. It could be difficult for the big IT companies to handle the information without mismanagement and abuse as a result [13]. Strong networking technology is essential for success due to the rising amount of sensors and their data. Such a concept is quite unlikely to succeed without robust citywide coverage. Millions of people will live in smart cities, making it difficult for AI and IoT to manage, analyse, and prevent cyber-attacks. In order to address data privacy, user security, and faith in smart cities, an integrated strategy is required. Meanwhile, all stakeholders participating in the smart city process, involving municipal managers, citizens, and mankind at large, must share the burden of information technology security and data threats and the accompanying obligations. A cyber assault is one of the different sorts of data threats and cyber security that are included. In order to develop sophisticated systems to identify threats, smart cities must implement effective safeguards. By outlining the constraints of monitoring,

5.6.3 Smart Hygiene and Waste Management

Increasing involvement has been shown in the possible use of these innovations to provide flexibility and efficiency in manufacturing processes and municipal services. Many nations still have serious concerns about how to handle trash. One of the biggest problems municipal corporations confront across the world is managing waste from creation to disposal. Handling the garbage created by millions of people is a significant problem given the present stage of living methods, in which the majority of eatables and other things are covered in plastic or paper cover. Cities produce garbage at an alarming rate, thus rubbish collection must be finished within a certain time limit. Cities also need to gather waste more shrewdly. Some waste management firms have created waste division methods which use artificial intelligence to autonomously separate various waste types (such as paper) without the need for human interaction. The problems related to waste management's collection, transportation, treatment, recycling, and disposal should be addressed via AI and the IoT. The intelligent disposal method, for instance, was made possible by ant colony optimization (ACO) technology. Smart city residents can receive high-quality services since the entire process can be centrally managed

5.6.4 Lack of Professionals

The IoT and AI adoption both need highly qualified individuals. Organizations frequently misinterpret the advantages of these technologies because they lack the necessary expertise and understanding. The scarcity of workers with expertise in computer technology and several sectors is one of the important issues that smart cities are experiencing. The use of AI and the IoT in the creation of sustainable smart cities is behind the lack of qualified experts. The shortage of specialists in Amsterdam and Hungary was thoroughly examined by Novak et al. In the post-pandemic age, attracting and adopting experts will be especially crucial. The city has already felt the effects of the epidemic's economic effects, and if there is a persistent shortage, it can become challenging to implement intelligent urban management which can have an influence on the effectiveness and value commitment of smart city operations. The expertise of experts aids in achieving desired outcomes for the creation of smart cities.

5.6.5 Managing Energy Demand

Advanced computing methods are incorporated into and transformed into common things to create smart gadgets, which are then used as intelligent terminals to exchange data with other devices or a cloud server. Thus, maintaining the smart city demands a significant quantity of energy. It may be difficult to supply a city with the energy it requires, and as more focus is being paid to renewable energy sources, some towns have found it difficult to make the transition. On the other hand, both the demand for and the price of energy are rising with time. Owing to the usage of contemporary gadgets, domestic energy consumption is exponentially rising. Higher energy demands are a result of both technological advancements and changes in consumer behaviour. As a result, energy producers are now looking to AI and the IoT for assistance in optimizing the arrangement of energy requirements through strategies like automatic functioning of streetlights, raising unit electricity bills during busy hours, and replacing outdated hard wares with more advanced techniques. AI recommends latest guidelines for scheduling tasks. Improved energy infrastructure design, implementation, and production are required to address a number of issues. Installing solar power-producing facilities for residences is one way to address the issue of excessive energy usage. Auxiliary solar power systems installed in contemporary residences can also contribute to increasing the value of fixed assets. After the sun sets, residents will start to utilize power from the major grid as customs because power is not produced by the solar panels at night.

5.6.6 Managing Transportation

A significant portion of the modern economy is the transport sector, which can contribute 6%–12% of a nation's GDP. Even though mobility has significantly enhanced our lives, there is still a great deal of unnecessary issues. Due to inefficiencies, the transport industry has grown to be the second-largest carbon emitter. Petroleum, diesel, and other products made from crude oil are significantly used in modern transportation technologies. Electric vehicles are an excellent substitute for helping with emissions-related issues. Electric motors and a battery pack in electric cars provide the required power for driving. If current gasoline stations are converted to hybrid models that offer both petroleum goods and e-vehicle charging stations, the need to charge electric vehicles may be addressed. It's crucial to plan a city to eliminate the need for everyday public transit that may fill in the gaps and shorten commute times for inhabitants in order to avoid traffic congestion. Smart transport is one of the major pillars of smart cities. Without a dependable and effective transit system, there can be no smart city. Because of this requirement, intelligent transport systems (ITSs) are an essential part of any smart city idea. This has an impact on both the environment and intelligent transportation. An instantaneous smart transport simulation for smart cities was created by Saroj et al. The viability of using network implementation measurements to deliver feedback in an effective world, as well as bulk data context, was assessed and visualized using this simulation model. The machine-learning strategies that are a key component of Artificial Intelligence should be able to analyse historical data from universal transportation ventures to identify the main reason of where there is frequently traffic congestion or where the majority of accidents occur, as well as potential solutions. A smart bicycle dispatch system, or shared bicycle, is available in China in addition to the smart taxi transmit system for cars. The algorithm suggests a sensible organizing plan based on current traffic circumstances, which will then be handled physically. This increases the effectiveness of shared bicycle use. For city dwellers, bicycle sharing is a "last mile" issue solution.

5.6.7 Environmental Risks

Climate change and environmental threats are making cities more susceptible. Extreme flooding that occurred in July 2021 in a number of European nations as well as China cost millions of dollars in property damage and claimed many lives. Smart cities should include extremely sensitive and adaptable disaster management systems, such as those that monitor the weather and warn residents about pollution prevention measures. The necessity to maintain economic expansion is continual due to the rising population, to the point that it poses environmental problems. Due to the vast population environmental issues like air pollution, pollution due to solid waste and disposal of non-reusable plastic have also increased by a considerable amount. Large-scale housing needs also pose several threats to the environment. In order to reduce environmental concerns, Artificial Intelligence and IoT may be employed at the design process. Automated drones have been utilized in a variety of industries, including air pollution monitoring, traffic control, and environmental risk detection and collection. Environmental dangers may be measured with the use of air quality sensors on openly available web platforms. Cities become smarter as a result of urban settings, which increase competitiveness and adapt to environmental threats.

5.6.8 Literacy and Healthcare Maintenance

By the year 2020, 86 enterprises involved in the healthcare industry would have invested around USD 55 million in AI and healthcare. It is difficult to provide healthcare facilities for everyone in China, which has the largest population in the world. Giving all age groups essential medical care without first determining their prior health status is tricky. AI should have the ability to evaluate a case's physical and past medical histories to determine their current state of health. Governments will be able to evaluate the data at any time, wherever in the nation, thanks to the digitization of medical information on a centralized server. This would enable medical professionals to properly examine the patients and treat them. In addition to keeping the medical records on a centralized server, explored how a blockchain may be used in the healthcare industry and identified it as a tool for decentralized healthcare administration. Kong suggested an AI-depend system for hospitals in smart cities. Operation rooms, diagnostic instruments, and other IoT-enabled medical equipment should help with health management. For instance, nations established steps for efficient protocols for exchanging health data during the COVID-19 pandemic. Better urban structural risk management choices, according to Allam, can only be achieved by creating various smart city outcomes to help standardized rules that enable smooth connections between them. The improper use of personal health data is the antithesis of modern technology's wise application. All data collecting and usage measures ultimately result in data violation or exploitation. For example, travel experiences related to the Corona pandemic or documentation of appointments for challenging health problems, that may jeopardise the morality of patients.

The residents' top priority also includes education. AI should aid in the planning and creation of educational initiatives that meet the demands of business and academia. One way AI may be used is in the ongoing improvement of curricula, evaluation of student abilities, and analysis of the industrial employment needs for students' skill advancement. In order to promote student involvement and connection with their classmates and professors, IoT may communicate and use information in extremely suitable ways.

5.6.9 Deficit of Assurance in AI and IoT

In a connection between a manufacturer and a customer, trust is a key component. The public's level of trust is rising in smart cities. The growth of smart cities may be slower if there is a lack of confidence in AI and IoT. Building a reliable AI model contributes to society's advancement and ensures a brighter future. Governmental organizations could work to raise awareness by disseminating information that supports the use of AI in sustainable practices, upholding openness, and sharing certain case studies that might increase public confidence. How AI may affect a typical city in 2030 was thoroughly examined in the paper "Artificial Intelligence and Life in 2030: One Hundred Year Study on Artificial Intelligence". The many effects and difficulties of AI are reflected in each of these chapters and the forecasts for the following 15 years. Examples include the difficulties of developing trustworthy and secure technology, the difficulty of earning the public's trust, and the dangers that less human connection poses to society as a whole. Poola noted both basic jobs that AI is capable of carrying out, like face recognition and driving a car, as well as sophisticated activities, like creating a super AI that evolves and causes an explosion in intellect.

It is also conceivable to deploy cutting-edge technology to end poverty while also creating sensible AI safety measures. An synopsis of AI and its uses in mankind was given by Kambleetal. For instance, they looked at how AI approaches are now used to defend computers and communication networks against cyber assaults. Technologies are utilized in the medical industry to enhance caring for inpatients in hospital. Data management systems offer the assembled records and produce income, but authorities must also do so while upholding the public's confidence. A major obstacle to the establishment of sustainable smart cities is managing and fostering public trust. Smart city governance that is focused on data and citizens is fundamentally founded on trust. Falco, on the other hand, put out the concept of AI and stated that involving the people is not enough to increase public confidence in AI. Hwang examined Korea's transition to smart cities. He claimed that because of their experiences with u-City, many Koreans had lost faith in the development of smart cities. However, the emergence of new technology has the potential to somewhat alter this circumstance. Building smart cities is impossible without the AEC (Architecture, Engineering, Construction) sector's input

5.7 ADVANTAGES OF SMART CITY

All facets of human activity, including science, governance, and health, will benefit from AI in the IoT. For instance, AI may significantly enhance marketing efforts in a competitive environment. Now, let's examine some benefits of AI in IoT.

5.7.1 Improves Operational Efficiency

Huge amounts of data are processed by AI, which detects similarities quicker and more correctly than humans. The IoT can leverage AI to deliver high-quality services and goods while consuming less time, money, and effort. AI uses a variety of techniques to increase productivity in businesses, one of which is data preparation. IoT devices capture both necessary and unnecessary data as they gather data. IoT AI may therefore sort through data to find what is desired before presenting illuminating facts for operational effectiveness.

IoT and AI enhance data labelling as well. In order for a machine-learning programme to comprehend and make use of information, labels or names must be added to raw information. The data type is described by these labels and tags. Data cleaning comes next after data labelling. The business may do this by fast-cleaning already-labelled data for inclusion in machine learning models using tools like MATLAB.

5.7.2 Predicts a Broad Range of Risks

Business issues are uncertainties or unforeseen occurrences that may be outside of the businessman's control and result in a loss. Businessmen and wealthy people may protect their data and money from criminals by using AI.

5.7.3 Creates New Products and Enhances Existing Ones

AI may make recommendations for developing novel items and aid in the modification of already existing ones. IoT, for example, can assist UI/UX design firms in automating chores and accelerating prototype creation.

In an effort to improve their company's manufacturing process and keep track of how their target market would engage with their product, product designers have also resorted to AI.

5.7.4 Predicts Equipment Failure

AI-enabled IoT sends information with precise standards for the amount of strain or pressure a piece of equipment is designed to sustain.

This can make it easier for businesses that use reactive maintenance to identify equipment breakdown and transition to preventative maintenance.

5.7.5 Helps Schedule Orderly Maintenance

The use of previous data by AI can assist large-scale manufacturers with many machines in scheduling maintenance in a systematic way.

5.7.6 Better Transportation Services

The existing state of transit in a city may be significantly improved by a smart city, improved traffic control, the capacity to track public transit, and improved service to its residents through continual information and affordable rates, which will all be features of this city.

5.7.7 Safer Communication

Technology can be most advanced in a smart city, and collaborations with the commercial sector can benefit society as crime will decrease. Such technology includes, for instance, body cameras, gunfire detectors, networked crime centres, and licence plate recognition.

5.7.8 Efficient Public Services

Smart cities will have the technology and tools required to reduce our use of natural resources and minimize waste of water, power, etc., without having to reduce any other variables, as there are only so many natural resources that can be used to fulfil the demand of the population.

5.7.9 Reduced Environmental Footprint

Thousands of energy-efficient structures may enhance air quality, utilize renewable energy sources, and reduce reliance on non-renewable energy sources in a smart city. The ecological influence we have on the environment can be lessened thanks to these.

5.7.10 More Digital Equity

People must have access to inexpensive high-speed Internet services and equipment. All city dwellers can have equal possibilities if they can access free Wi-Fi in neighbourhood hotspots.

5.8 DISADVANTAGES OF SMART CITY

Smart City has lots of advantages, but there are also some disadvantages. The disadvantages are as follows.

5.8.1 Limited Privacy

The citizens may struggle to keep their anonymity since the authorities or the government will have access to security cameras and sophisticated systems connected through many different areas. The idea of privacy or personal space will be radically altered by facial recognition technology and similar technologies.

5.8.2 Social Control

Greater power may go to those who can track and centralize the information that security cameras collect. It may be the government, a business, or some other kind of authority. They can be able to effortlessly sway public opinion and have control over a citizen's data.

5.8.3 Excess Network Trust

The inhabitants of these smart cities might lose their autonomy in decision-making and may even become inept since they will rely nearly totally on electronics and networks. If these tools are not available, they will not be able to respond correctly.

5.8.4 Difficulty in the Pre-Commerce Stage

The smart technologies might still be in their pre-commercial phases even though money will be available. These cities won't have the technological expertise or resources.

5.8.5 Pre-Training is Required

The inhabitants of the city won't be able to employ technology if they don't understand it. Without instruction, students may find it impossible to apply and irrelevant to their daily lives.

5.8.6 Security Concerns

IoT and AI provide better security measures, but they also introduce some vulnerabilities. Threats to cyber security exist for both new and existing enterprises. Increased connection, especially through IoT, may lead to data breaches, data loss, and hardware- and network-based attacks. The next largest problem is technical complexity. Even simple system programming changes need the hiring of skilled developers, programmers, and data scientists by organizations. There are many complexities involved that are challenging for an average employee to understand, let alone solve.

5.8.7 Power Dependency

Your system as a whole is powered. You utilize power to connect and run the things you use. So, your entire network may go down at any time if you experience a power outage or a software fault. Your company might experience significant losses. There can be a lot happening in the fields of AI and IoT in 2022 that might make integration efficient and smooth. There will be steps done to increase security and close all gaps. The system may also develop to provide less complexity and higher scalability. The sectors might be dominated by trends including the expansion of edge computing, cutting-edge wearable technology, and cutting-edge automation. So, keep up to date to boost your business.

5.9 CONCLUSION

This chapter covered a wide range of topics related to smart cities and the Internet of Things. Giving clear idea of various improvements and numerous technologies that are implemented in the smart cities, we highlight IoT as an important enabler of smart city concepts, talk about the various smart city building, and address the difficulties encountered while deploying smart city utilizations. The use of AI in smart cities is then debated after an analysis of the networking and sensing mechanism employed for similar approaches. In order to provide a survey of the recent research plans in Smart Cities with IoT, we have carefully considered the kind of deployment for every implementation presented for the different elements. This decision is based on the technologies and architectures discussed. This survey is intended to help academics by offering a thorough initial stage for the Smart Cities where IoT is used. This chapter suggests that for the purpose of better function, smart cities must adopt artificial intelligence technologies. In addition to adopting artificial intelligence, smart cities should implement plans to integrate the technology into the many operations needed to build a municipality. Such actions can guarantee that the idea of smart cities is acknowledged and accepted in the majority of the world. In conclusion, the creation of intelligent cities has the promise of enhancing both the public and private sectors' standard of living in metropolitan areas. To increase their sustainability, smart city development must take technology into careful consideration. In order to increase security, smart policing, correct waste management, efficient energy usage, and smart parking, artificial intelligence is essential in smart cities. With the aid of installed cameras and IoT-capable sensors, AI gathers information from data. The appropriate bodies are then given access to the intellect so they may make educated wisdom. Although some environmentalists disagree with urbanization, AI makes it more advantageous.

REFERENCES

1. Alahi, M.E.E., Sukkuea, A., Tina, F.W., Nag, A., Kurdthongmee, W., Suwannarat, K., Mukhopadhyay, S.C. 2023. Integration of IoT-enabled technologies and artificial intelligence (AI) for smart city scenario: Recent advancements and future trends. *Sensors* 23(11): 5206.
2. Syed, A.S., Sierra-Sosa, D., Kumar, A., Elmaghraby, A. 2021. IoT in smart cities: A survey of technologies, practices and challenges. *Smart Cities* 4(2): 429–475.
3. Li, Z., Shahidehpour, M., Bahramirad, S., Khodaei, A. 2017. Optimizing traffic signal settings in smart cities. *IEEE Trans. Smart Grid* 8(5): 2382–2393.
4. Sivasankar, B.B. 2016. loT based traffic monitoring using Raspberry Pi. *Int. J. Res. Eng. Sci. Technol.* 1(7): 2454–664.
5. Narendran, S., Pradeep, P., Ramesh, M.V. 2017. An Internet of Things (IoT) based sustainable water management. In *Proceedings of the 2017 IEEE Global Humanitarian Technology Conference (GHTC),* San Jose, CA, USA., pp. 1–6.
6. Lowe, M., Qin, R., Mao, X. 2022. A review on machine learning, artificial intelligence, and smart technology in water treatment and monitoring. *Water* 14: 1384.
7. Anagnostopoulos, T., Zaslavsky, A., Kolomvatsos, K., Medvedev, A., Amirian, P., Morley, J., Hadjieftymiades, S. 2017. Challenges and opportunities of waste management in IoT-enabled smart cities: A survey. *IEEE Trans. Sustain. Comput.* 2(3): 275–289.
8. Kabir, M.H., Hasan, K.F., Hasan, M.K., Ansari, K. 2021. Explainable artificial intelligence for smart city application: A secure and trusted platform. arXiv 2021, arXiv:2111.00601.

9. Borgia, E. 2014. The Internet of Things vision: Key features applications and open issues. *Comp. Commun.* 54: 1–31.
10. Lv, Z., Qiao, L., Kumar Singh, A., Wang, Q. 2021. AI-empowered IoT security for smart cities. *ACM Trans. Int. Technol.* 21: 1–21.
11. Nikitas, A., Michalakopoulou, K., Njoya, E.T., Karampatzakis, D. 2020. Artificial Intelligence, Transport and the Smart City: Definitions and Dimensions of a New Mobility Era. Sustainability
12. Ismagilova, E., Hughes, L., Dwivedi, Y.K., Raman, K.R. 2019. Smart cities: Advances in research: An information systems perspective. *Int. J. Inf. Manage.* 47: 88–100.
13. Haikh, Y., Parvati, V.K., Biradar, S.R. 2018. Survey of smart healthcare systems using internet of things (IoT). In *2018 International Conference on Communication, Computing and Internet of Things (IC3IoT),* pp. 508– 513, Chennai, India.

6 Securing Smart Cities
Addressing Cyber Security Implications and Collaborative Measures

M. Kalaiyarasi, S. Karthi, K. Kavya, V. Karthika, and Smita Sharma

6.1 INTRODUCTION

The main aim of smart city is the quality of life should be improved and the city function. Smart cities use Information and Communication technologies (ICT) to improve performance. The smart city focus on the areas like health, education, and transportation. The determination of smart cities is by better infrastructure, good public transportation, intelligent shopping, smart street lights, electric vehicle running, and Smart cities will also be called as "Intelligent community" and "smart community". Smart cities have created great attention in recent years. It has created lot of advantages to the people. It will create many job offers and better connectivity. The father of smart city is Walt Disney. It will help the patient's growth. Singapore is considered as the global smart cities. The smart city has some issues like the privacy is limited, it has worry of data privacy and implementation is difficult. The Ways that citizens should know about the thing is citizen should be knowledgeable. The citizens must have equal access to all new technology for better living [1]. All the cities can use the concept of smart city and increasing the everyday life in better way. Smart people will create the smart cities in good way. The youngster must have to put effect on the smart city to upgrade the future.

Cyber security refers to unauthorized users attacking data and programme, and it is known as a cyber-attack. The protection of this attack is called cyber security. Cyberattacks are committed by cybercriminals using smart devices. The attack is dangerous to the user. Cyber security is protecting our data from malicious activities. Keeping them safe is important because nowadays, we use smartphones and gadgets, and data are stolen by attackers. The ways to prevent cyber-attacks are by upgrading the software systems and, not opening unwanted links, verifying the email, the data should be backed up. The passwords should be different for different platforms; having the same set of passwords will be extremely dangerous to the user. Use a firewall for protection from hackers. One type of cyber security threat malware; when we click on the link, it will create dangers, and phishing leads to fake communication such as opening an email and getting the information and denial of service, a man in

DOI: 10.1201/9781003459835-6

the middle, Emotet, SQL injection, password attacks are some common cyber security threats. Cyber attackers steal the company's information, so that's why cybersecurity is needed to keep the company's data private. The different types of cyber security are network security. There may be a lot of advantages in cyber security such as protecting the important data, detecting the attacks earlier and minimizing damage. All should keep our data and devices private and secure. There are some disadvantages in cyber security implementation cost is high, management is difficult, and effectiveness is limited. Some principles of cyber security are secure configuration risk management, network security, create awareness among the user. Secure configuration means making the devices secure to reduce attacks. Risk management is recognizing and applying steps to get prevention. Malware prevention involves using antivirus programmes to prevent harmful software. In the current world, cyber security is important for all people. Having cyber security can decrease the number of harmful attacks from users. Network security means control of data loss, Identity access management, and Network access control. Cyber security means protecting the cloud devices [2]. Mobile security secures the devices from jail breaking. Application security can prevent the top ten attacks and stop vulnerable actions. Zero trust includes providing security and protecting individual resources.

Cyber security is important for smart cities because, in smart cities, we share and manage a large amount of data; to protect this type of data, cyber security is important. Smart cities are greatly affected by cyber-attacks in different ways. Nowadays, new technologies have been introduced to increase the quality of citizen life. So the use of new technology will lead to some problems. To prevent this problem, cyber security is used. In smart cities, any malicious activities done by an individual or some organization can put the whole world at risk. Smart city architecture includes smart energy, smart buildings, smart healthcare, smart security, smart environment, and smart public services. In smart cities, data is collected and transmitted to the government. No technology is completely safe, and it may cause illegal activities. So, to prevent this, cyber security is used. Advanced Persistent Attack is the most common form of cyber-attack in smart cities. Hackers extract the data from the public. Ransomware is an attack in smart city. Ransomware scans the local devices and affect the devices. Organizations should implement with cyber security requirements to know about protection of private data and security. The cyber-attacks against smart cities leads to financial loss, loss of private data disturbance in infrastructures.

There are lots of attacks in smart cities so to prevent cyber security is needed. Maintaining data from attacks and bad actions is the duty of cyber security. The first step in cyber security is finding the problem in cities and cyber security challenges. Without analysing these types of things, we cannot give perfect solutions for the city. Deep learning is used by the researchers. Deep learning continuously analyse and gathers information to help the system. Innovative technologies will lead cities to make better and improve citizen's life [3]. Smart cities include sensors like smart parking sensors and Internet of Things sensors; implementing smart cities is very difficult because hackers are more creative nowadays, so to avoid the difficulties, cyber security is used. Security risks are increasing nowadays. Protecting our city and making a happy life is important. It's all under the control of citizens. Every citizen must have a great effect on the smart city. Citizens need to trust the security of smart services. Smart system services create greater benefits for the citizens. With the proper support and guidelines, everyone can make the smart city better.

6.2 SMART CITY

A smart city (SC) is an urban area that uses data and technology to enhance the quality of life for its citizens. This city tackles urban problems and improves many facets of city life, including safety for everyone, transportation, electricity, waste management, and government, by using creative solutions and connections. It collects and analyses data in real time through the use of sensors, networked devices, and computerized analytics, which facilitates improved decision-making and effective resource allocation. The Internet of Things (IoT), which links devices and gathers data on various aspects of urban infrastructure, such as waste management, energy usage, and transport, is an essential piece of an intelligent municipality. Making educated decisions is made possible by processing and insight extraction of the gathered data using analytics of data. By incorporating green technology, such as energy-efficient construction, smart grids, renewable energy sources, and effective transit systems, smart cities also prioritize sustainability [4]. Utilizing smartphone applications, web portals, and social media platforms, they place a high priority on public involvement and include locals in decision-making processes. By putting intelligent transportation systems, real-time public transit developments, and innovative parking solutions into place, smart cities want to increase mobility. By implementing cutting-edge surveillance systems, video analytics, and emergency response mechanisms, they also place a high priority on safety and security. Enhanced utilization of resources, less of an adverse effect on the environment, better quality of life, more economic possibilities, and better governance are some advantages of smart cities. However, for the implementation to accomplish the desired outcomes, it needs to deal with challenges including concerns regarding privacy, data security risks, and equitable access to services and technological resources.

The idea of the New World Order (NGO) started to acquire traction in the late 20th and early 21st centuries as industry and technology advanced rapidly. The phrase "smart city" gained popularity during the period when consumers could finally purchase and access new technology on a large scale. The concept of utilizing technology to live a sophisticated and contemporary urban life has been around for a while, as seen by the sheer volume of projects and initiatives that are being undertaken around the urban globe [5]. A turning moment in the development of smart city technology and methods was the establishment of the SCouncil in 2012, a group tasked with creating a campaign. Additionally, the United for Smart Sustainable Cities (U4SSC) smart sustainable cities initiative was established in 2016 in order for the UN to assist cities in the process of developing the smart and sustainable projects. Some communities have adopted the concept of a "smart city" throughout time, and innovation in this area has been becoming better. Depending on the demands and resources available in each location, the designs and features of the various smart city initiatives may need to vary even within the national framework.

SCs are intended to address the complex challenges that industrialization is producing, such as air pollution and traffic, as well as the cutting-edge technology that may be used to enhance urban living. The concept of SCs came forth as a result of the problems that cities are facing: a lack of resources, deteriorating environmental conditions, inadequate infrastructure, and high resident expectations [6]. Modern cities use data analytics, communication technology, and developed technologies to improve resource utilization, address social issues, and improve the standard of living for their citizens. The goal of smart cities is to support the mayor's policies, which include inclusion and diversity along with innovation, resilience, and economic growth. In an effort to address urban problems, enhance liveability, and secure the future of future generations, SCs make use of evidence-based decision making, effective resource management, and public participation.

6.2.1 Types of Smart Cities

It is feasible to create a variety of smart cities, each with a distinct emphasis on efforts and level of technological integration. The following are typical types of smart cities: The following are typical types of smart cities:

6.2.1.1 Smart Cities Propelled by Technology

The technological environment in smart cities uses the most recent methods, which increase sustainability, efficiency, and people's well-being. Urban innovation pushes cities to create networked, effective systems that save money and enable data-driven decision-making by municipal leaders. In SCs, where technology serves as an integrator, sensors and Web-enabled devices make up a significant portion of the building or urban infrastructure. These sensors record data on a variety of characteristics, including water usage, disposal, utilization of energy, traffic movement, and air quality, all in real-time [7]. AI systems and sophisticated algorithms are then used to process and evaluate the gathered data in order to extract insightful information and make defensible conclusions. Intelligent cities powered by technology prioritize enhancements in the following areas: transportation, energy efficiency, safety and security, citizen services, urban development, and the preservation of the environment. Generally speaking, technologically driven smart cities use data analytics and cutting-edge technologies to improve urban operations, raise living standards, and encourage sustainable growth. The objective of these cities is to create environmentally conscious, effective, and linked urban settings by means of efficient integration and usage of technology.

6.2.1.2 Smart Towns with an Emphasis on Mobility

Enhancing mobility and transit inside cities is a top priority for mobility-focused SCs. These cities make use of cutting-edge technology and creative strategies to improve convenience, environmental responsibility, and effectiveness [8]. Creating smooth, practical, and eco-friendly mobility choices for locals and guests is the aim. Intelligent Transportation Systems (ITS), Public Mobility Services, Active Transportation, Connectivity and autonomously vehicles, and mobile technology as a service are important components of mobility-focused cities that are smart. Smart cities that prioritize accessibility seek to eliminate traffic jams, increase accessibility, improve air quality, and promote sustainable transportation options via the use of technology and creative thinking.

6.2.1.3 Smart Cities Focused on the Citizenry

Citizen-centric SCs put inhabitants at the heart of urban growth and processes for making decisions, giving residents' interests and well-being first priority. These cities work hard to include their residents, provide them with technological resources and information, and guarantee their involvement in the future development of the city [9]. These cities seek to foster a feeling of citizen possession, engagement, and cooperation by using citizen-centric practices. They encourage a strong feeling of community, empower people, and advance accountability and openness. In the end, citizen-centric smart cities put the wants and goals of its citizens at the centre of urban development in an effort to improve their quality of life and overall well-being.

6.2.1.4 Safety and Security-Oriented Smart Cities

Protection and security ecosystems (SCs), which are made possible in metropolitan areas by the use of data-driven procedures, technologies, and other cutting-edge concepts, exist primarily for security reasons. Their primary objective is to use proactive, reactive, preventive, and predictive techniques to address unanticipated safety and security challenges. Their goal is to create a robust and secure environment that benefits locals, businesses, and visitors alike [10]. Typical characteristics and programmes of safety and security-focused smart cities include the following: Cyber regulations public safety apps, proactive law enforcement, smart traffic management, emergency response mechanisms, smart surveillance systems, and community engagement and awareness. After all, smart cities prioritize security and safety, manage data analytics via the use of cutting-edge technology, collaborate with partners, and create resilient communities that meet high safety requirements. These cities want to lower crime rates, improve emergency response times, and give their citizens a secure place to live by combining these strategies.

6.2.1.5 Sustainability-Focused Smart Cities

Urban regions that put a priority on conserving resources, sustainable development, and the general well-being of both people and the natural environment are known as sustainability-focused SCs [11]. These cities use cutting-edge tactics and technology to lower their carbon footprint, encourage growth that is environmentally friendly, and improve the quality of life (QoL) for their citizens. Key features and programmes that are frequently observed in smart cities with a sustainability component are as follows: Water management and conservation, data-driven sustainability, waste management and recycling, green infrastructure and urban planning, renewable energy generation, energy efficiency measures, sustainable transportation systems, and public awareness and education. These cities want to improve the quality of life, lessen their influence on the environment, and increase their resistance to climate change by emphasizing sustainability. They operate as role models for environmentally friendly urban growth, encouraging other cities to follow suit in the sake of a more sustainable future.

6.2.1.6 Digital Smart Cities

Smart cities and technological infrastructure, together referred to as digital intelligent cities, aim to enhance quality of life, promote sustainability, and streamline city processes [5]. These cities collect and analyse data in real-time through a network of linked devices, sensors, and data analytics, which facilitates improved decision-making and effective allocation of resources. Key components of digital SCs include Internet of Things (IoT), Data Analytics, Connectivity, Sustainable Infrastructure, Citizen Engagement, Smart Mobility, Safety and Security. The benefits of digital smart cities include improved resource efficiency, reduced environmental impact, enhanced quality of life, increased economic opportunities, and improved governance. However, challenges like privacy concerns, data security risks, and equitable access to technology and services need to be addressed for successful implementation.

6.2.1.7 Resilient Smart Cities

Resilient SCs are urban areas that not only leverage digital techniques but also prioritize resilience in the face of various challenges, including natural disasters, climate change, social disruptions, and economic shocks. These cities integrate advanced technologies, data-driven solutions, and robust infrastructure to enhance their ability to withstand and recover from these challenges efficiently and effectively. Here are key aspects and characteristics of resilient smart cities: Risk Mitigation, Adaptive Infrastructure, Data-Driven Decision Making, Collaboration and Engagement, Smart Infrastructure Monitoring, Integrated Systems, Social Resilience, Social Resilience [12]. The benefits of resilient smart cities include reduced losses from disasters, improved emergency response, improved public safety, increased economic stability, and better QoL for residents. By combining resilience strategies with smart city technologies, these cities can effectively respond to challenges and adapt to changing circumstances, making them more sustainable and liveable in the long term.

6.2.2 Components of Smart City

SCs are defined by the hybridization of various technologies and infrastructure components to enhance the QoL for residents, improve sustainability, and streamline city operations. While the specific components can vary from one smart city to another, here are some major components commonly found in smart city initiatives, as shown in Figure 6.1.

IoT Devices: These are interconnected devices embedded with sensors, actuators, and communication capabilities. IoT devices, such as smart electricity meters, garbage collection systems, traffic detectors, and air quality monitors, collect data in real time from the surroundings.

Data Mining and Large-scale Data: To analyse and make sense of the vast volumes of data gathered from various sensors, smart cities rely on big data technology and sophisticated analytics. This aids in the comprehension of patterns, the recognition of trends, and the formulation of data-driven choices for the best use of municipal resources and services.

FIGURE 6.1 Main components of Smart City.

Effective Resource Administration: In order to maximize consumption and improve sustainability, SCs incorporate technology for smart energy usage, including renewable energy sources of information, intelligent grids, energy-efficient structures, and sophisticated management systems.

Intelligent Structures: Structures with robotics, energy efficiency, and Internet of Things (IoT) devices are capable of tracking occupancy, optimize energy use, improve security, and make spaces more comfortable for its inhabitants.

E-Government and Citizen Services: Online resources and digital platforms make it possible for citizens and the government to communicate effectively and transparently. This covers digital identity systems, e-voting, online government service portals, and smart governance programmes.

It is crucial to remember that elements of smart cities are interrelated and function as a whole to enhance efficiency, sustainability, and urban life [13]. The particular combination of elements employed in a smart city will vary based on its objectives, top priorities, and funding that is accessible.

6.2.3 Main Aim of Smart City

In the modern era, a SC's primary goal is to maximize the power of innovation and technology. Smart cities use data and digital technology to improve infrastructure and services, limit environmental impact, increase economic growth, and improve the quality of life for their citizens. These cities collect and interpret vast amounts of data through the use of networked sensors, devices, and networks, which improves resource management and decision-making. Transportation, energy, waste management, public safety, healthcare, education, and governance are among a smart city's primary priority areas. Enhancing citizen involvement, allocating resources optimally, lowering pollution and traffic, enhancing public services, and building inclusive communities are the goals of SCs.

Building on current developments and utilizing new technology, a SC of the future will create urban settings that are even more resilient, sustainable and focused on the needs of their citizens. Figure 6.2 graph illustrates the present state of smart city trends. The future's socially connected HR professionals (SCs) will, nonetheless, be far more understanding, perceptive, and connected. First and foremost, the primary goal will be to improve QoL by utilizing contemporary technologies. Future smart cities will prioritize environmental preservation above all else, offering its residents clean and effective waste management techniques, a large-scale reliance on renewable energy sources, and an ecologically friendly footprint. The integration of smart grids, energy-efficient buildings, and modern transit systems into architectural designs is necessary to achieve resource efficiency and promote environmentally conscious living.

Smart cities' advanced transportation infrastructure will primarily comprise self-driving automobiles, intelligent traffic control, and shared mobility, public transportation, and active transportation options like cycling and walking. The goal of pedestrianization is to lessen traffic, facilitate better access between locations, and give locals unrestricted use of public transportation. Furthermore, the involvement of citizens need to be the primary goal of the future generation of smart cities. They want to use technology to promote inclusive and transparent administration, hence promoting citizen participation in the making of decisions. Real-time input, involvement, and cooperation will be made possible via digital platforms, enabling

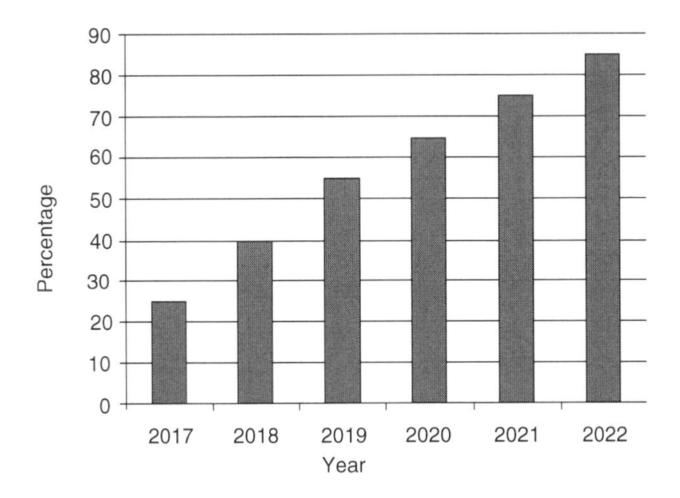

FIGURE 6.2 Smart City development.

locals to influence the growth and betterment of their neighbourhoods. In addition, the future's intelligent cities will transform medicine [14]. Advanced analytics and AI algorithms will facilitate early disease detection, preventive care, and efficient resource allocation in healthcare facilities. SCs in future generations is to create intelligent and sustainable urban ecosystems that prioritize the well-being of residents, enhance liveability, foster innovation, and adapt to the evolving needs and challenges of a rapidly changing world.

6.3 CYBER SECURITIES

Cyber security means protecting devices and employees from cyber threats. Cyber-attacks are common nowadays, so to avoid cyber-attacks, cyber security is implemented. There are various types of cybersecurity solutions that can be used to prevent cyber-attacks. In the current world, our life is fully dependent on the technology. With some new technologies, it will create hard problems. The cyber-attacks mainly focused on accessing and devastating the information from users, robbing money from people, and disturbing businesses. Cyber security is very important to protect data from cyber criminals. The work of cyber security is by it includes new technologies and methods and analysis the problem to secure data from cyber-attacks. In 2011 Sony network get suffer from cyber-attacks and in 2020 many Facebook profiles were went for sale. Different countries have different government for regulate the cyber security. The cyber-attacks are common so awareness to the people is important. There are some techniques to get prevent from cyber-attack such as update the all hardware, software and operating systems, Do not open the unwanted pink which may cause harmful action from cyber-attack, Same types of password for different platforms leads to severe attacks, so using different passwords for different field's Everyone should have to back up their data from a removable hard drive. Everyone should test their network to see if any malicious and security risks happen to unauthorized users.

There are some common controls in cyber security they are Preventive control includes an action that should not allowed to happen and accessing Recovery control means if the data is loss, we can get back from this control. Logical control means it will limit access from hardware or software such as fingerprint readers and encryption. Physical control means it will limit access in physical ways used to prevent unauthorized access to sensitive data. Some examples of physical controls are security guards, thermal alarm systems, and biometrics. Security control includes antivirus software, passwords, firewalls, and two-factor authentication. Operational security includes decision-making to ensure that data and resources are secured. Application security control includes protecting devices from threats. Management control focused on risk management and information security management.

Operational control includes the processes that are executed by people and implemented by technical and are saved by the operational control. Detective control includes log monitoring, SIEM, security audits, trend analysis, video surveillance, and motion detection.log monitoring means it analyses real-time data. SIEM means tools and services are used to predict various systems operations. Trend analysis means identifying patterns to obtain information. Video surveillance digital videos and images are monitored. Motion detection means motions are detected using sensors. Corrective control includes backups and system recovery. Backups and system recovery means it will create and store the data copies, and it can be used to data backup.ips will prevent malicious activities. Different controls include cable locks, hardware locks and video surveillance, and compensating means it will satisfy the needs of security. The time-based OTP is the best example for compensates control. When the password is generated, it will be used for a few minutes after the time is over. We cannot use a certain password.

6.3.1 Types of Cyber Attacks

6.3.1.1 Malware

Malware is malicious software that leads to hard damage to networks, computer systems and hardware devices. Malware programmes are designed to steal information and money from users. There are two different types of malware: viruses and Trojan horses. First, viruses are different types of vulnerable programmes that, on execution, replicate themselves, and different files and programmes attach. If the attachment is successful, it means the targeted programme gets affected by the computer virus. Trojan horse means it is looks like exact software but it will harm the installation. The different types of viruses are booting sector virus, macro virus and direct-action viruses. Boot sector virus means it infects the master bot. A macro virus is a virus that places on the document-related data and infects the system. Direct action means viruses get attached by files, activate the virus programme and damage the systems. Different types of Trojan horses are Back door Trojan, cryxos Trojan, and Ransom Trojan. Back door Trojan creates the backdoor and remote access to the hacker, and the cyber criminals will affect the system. Cryxos Trojan horse means it enters the system by clicking some unknown messages and hacker can access the system. Ransom Trojan means it blocks the users from accessing its own systems.

6.3.1.2 Preventing Malware Attack

- The way to prevent viruses is by using certified antivirus software. The safe websites should be visited. The way to protect Trojan Horse is by avoiding clicking on unwanted messages.
- We can use a firewall to protect from attack.
- Keep the software and hardware devices updated.
- Be aware of opening the email. File sharing should be limited.

6.3.1.3 Phishing Attack

A phishing attack means a deceptive message is sent to the target user, and it is a basic trick to revealing important information from users like passwords. An example of phishing is when a person receives an email to update a credit card and clicks on that link. The hacker will use a trick on that link, and the money or information will be missed by the person. The phishing attack is done through email, messages, and communication. During communication with the user the hacker will steal the information. Pharming attack means some malicious programmes are made to create fake websites to steal the data. Spear phishing will acquire location through online messaging. The prevention from phishing attacks is as follows,

- Installing firewalls can prevent cyber-attacks. We can use desktop firewalls and network firewalls to protect, and they will give security.
- We have to update all the services; if we ignore the update, there could be a high chance of a phishing attack.
- Don't click on unwanted links through email or messages. Some fake websites can steal information.
- Don't give the information to the unsecured website; if the websites do not start with "http" or security certificates, avoid those websites.

6.3.1.4 Man in the Middle Attack

It is a cyber-attack when a communication between two parties a man in the middle will use some hack to get the information. The attacker can interrupt and control all the information. It is a threat to online security; the attacker can capture important information such as account details and login credentials. It can also be called a monster in the middle, a machine in the middle, a monkey in the middle, Man in the browser. Man in the browser mainly focuses on infection on the browser. Man in the middle is malware is distributed so the attacker can easily access the data. The types of Man in the middle attacks are Internet protocol spoofing, Domain name spoofing, HTTP spoofing, sockets layer hijacking, email hijacking, Wi-Fi dropping, session hijacking cache poisoning. Email hijacking means an attacker can gain access to the mail accounts of the users. The domain name system spoofing means users think they have reached a safe website, but they don't know that it are operated by cyber criminals Examples of Man in the middle attack were super fish, Diginotar.

6.3.1.5 Some Preventive Methods of Man in the Middle Attack

- Communication should be secured because the third party will see our activities on any time.
- Avoid using public websites and use the safe websites.
- Use the Virtual private Network (VPN) and the connections should be secured.
- The best way to stop email hijacking by enable the two factors authentication.

6.3.2 Types of Cyber Security

6.3.2.1 Network Security

It is used to secure the network and data. It has both hardware and software technologies. It mainly focusses on threats and trying to avoid spreading. Network security combines multiple layers in the network only authorized user can get access to network. The Network security is working by checking the right access and determine the access levels. The types of network security are physical, technical, administrative. Physical level is the basic level it includes protecting data and network from unauthorized users. Technical level includes it mainly focused on safeguarding the data. It has two functions: one depends against unauthorized users, and the other is defence against vulnerable actions. The benefits of network security are keeping the data safe and protecting against cyber-attacks [15]. Types of network security are firewalls, access control, and encryption. The Network security will protect against viruses, worms Trojan, and Spyware. Network security is very important because it prevents data from being stolen by cybercriminals. The model for network security is first we have to design the algorithms, secrete information should be generated, design different processes for sharing the information and protocol should specify. The benefits of network security are protection from internal and external threats, keeping the data safe, and avoiding risks.

6.3.2.2 Mobile Security

Mobile security means protecting the information from laptops, mobile phones, and tablets. The main advantages of mobile security are data backup and remote access. The importance of mobile security is that passwords should be protected, public Wi-Fi should be avoided, and apps should be conscious of them. The threats in mobile security are using untrusted mobile devices, using websites created by unknown persons and interaction with other systems. To avoid this threat, there are some elements in mobile security they are device security, Traffic security, and Barrier security. Device security means the devices which we are using Bluetooth, laptops, and mobile phones. In this, we should protect passwords and update the systems.

6.3.2.3 Application Security

Application security means testing the security features and to prevent from threats. It can test and tells the weakness of the application, and we can prevent attacks. The types of application security are authentication, encryption, testing, and logging. Application security in the cloud poses extra challenges. Special care has to be taken if the user only wants to access the data. Sensitive data is unsafe in cloud applications; in this data, transfers from users to applications. The main objectives of application security are protecting the code and fighting against attacks. We can apply application security design and development.

6.3.2.4 Cloud Security

Cloud security secures cloud systems. It will save the information from infrastructure and set up the cloud. The challenges in Cloud security are lack of visibility, multinancy, compliance, and mis-configuration. The types of cloud security solutions are Identity and access management, data loss prevention, security information and event management (SIEM). The National Institute of Standards and Technology (NIST) has made a practice to secure cloud computing. The companies continuously migrate the data to the cloud, and it will be unsafe for the third party to access the data. Cloud security is composed of data security, Identity and access management, data retention, and legal compliance. The cloud security is designed by physical networks, data storage, applications, data servers, and Operating Systems. There are two main components in cloud security: cloud service types and cloud environments. The work of cloud security is to protect storage, detect errors, reduce impact on the system, and enable data recovery. The Cloud security risks are internal threats by error and external threats.

6.3.2.5 IoT Security

IoT Security means Internet of Things security, which secures the Internet-connected devices and network. The types of IoT Security are Network security, embedded, and Firm ware assessment. The industries need IoT security to secure their industries. IOT securities are used in consumer application and government application. The Internet of Things is managed internally and externally. Command and Control centre are responses to internal such as software devices. IoT devices are classified into two types they are General devices and sensing devices. Home appliances are examples of general devices. IoT devices are used in smart watch, home, car, smart city, and healthcare. The levels of IoT security are device, edge, fog, and cloud [16]. The layers in IoT architecture are perception, transport, edge, processing, application, business, and security layers [17].

6.4 CYBERSECURITY AND PROTECTION FRAMEWORK

A framework for cybersecurity and protection is a structured method to handle and reduce cybersecurity risks within an organization. It offers a thorough set of guidelines, best practices, and controls to safeguard information systems, data, and assets from unauthorized access, use, disclosure, disruption, modification, or destruction. There are several cyber security frameworks available, each with its own set of principles and guidelines. Here are a few widely recognized frameworks:

NIST Cybersecurity Framework (CSF): Developed by the National Institute of Standards and Technology, the CSF provides a risk-based approach to managing cybersecurity risks. Organizations may evaluate and strengthen their protection of cybersecurity with the use of this approach.

CIS Controls: You ought to abide by this collection of cybersecurity best practices. To serve as a basis for creating an information security programme, it offers a list of the top 20 crucial security measures, arranged according to criticality. Even though they are just a few of the security subjects that are related to the controls, access control, asset management, vulnerability management, and incident response cover many important areas.

Information and Related Technologies Control Objectives: The Information Systems Audit and Control Association (ISACA), which focuses on governance in relation to IT system security, is credited with creating the COBIT framework. It provides a framework for connecting organizational objectives with IT initiatives and guidelines for managing cybersecurity risks. The following crucial steps are often taken by organizations when implementing a cybersecurity strategy,

Organizing: Strategy is security that is applied using the acquired template. Make budgets and targets, then plan an implementation timeline.

Execution: Specifically, carry out the following tasks: create policies and procedures, provide staff training, install safety precautions, and facilitate recovery and reaction plans.

Observation and Continuous Formation: Regular security monitoring, accurate risk assessment, and recurring audits of the efficacy of implemented controls are all important.

6.5 CONCLUSION

To assure the security of digital systems infrastructure, data, and networks, each SC must establish a cybersecurity and protection strategy. For heavily networked smart cities, this is a major cause for worry. Therefore, it's critical to implement the required safeguards to get rid of such threats and shield private information from prying eyes. In order for smart cities to successfully confront challenges in the future, the framework may be continuously evaluated and developed in addition to taking the necessary security measures against cyber-attacks. It takes a lot of time and effort to carry out this procedure on a regular basis. Prioritizing cybersecurity is necessary for effective deployment and operation in highly technologically advanced urban environments where technology-dependent governance is still a top concern. The neighbourhood of the future becomes more strong and safe when it interacts with the advantages of digital transformation and new innovation in smart city programmes.

REFERENCES

1. Ashokkumar, L., Jayasree, L.S., & Manimegalai, R. 2019. *Proceedings of International Conference on Smart Grid and Smart City Applications: Jobs in Smart City, Improvement in Life.*
2. Pallathadka, H., Mustafa, M., Sanchez, D.T., Sajja, G.S., Gour, S., & Naved, M. 2023. Impact of machine learning on management, healthcare and agriculture. *Materials Today: Proceedings* 80: 2803–2806.
3. Chong, J.L., Chew, K.W., Peter, A.P., Ting, H.Y., & Show, P.L. 2023. Internet of Things (IoT)-based environmental monitoring and control system for home-based mushroom cultivation. *Biosensors* 13(1): 98.
4. Alahi, M.E., Sukkuea, A., Tina, F.W., Nag, A. et al. 2023. Integration of IoT-enabled technologies and artificial intelligence (AI) for smart city scenario: Recent advancements and future trends. *Sensors* 23: 202–212.
5. Navarathna, P.J., & Malagi, V.P. 2018. Artificial intelligence in smart city analysis. In *2018 International Conference on Smart Systems and Inventive Technology (ICSSIT)*, pp. 44–47.
6. Räty, T.D. 2010. Survey on contemporary remote surveillance systems for public safety. *IEEE Transactions on Systems, Man, and Cybernetics, Part C (Applications and Reviews)* 40(5): 493–515.
7. Portmann, E.R. 2015. *Smart Cities: Big Data, Civic Hackers, and the Quest for a New Utopia*, W W Norton & Company, New York/London.
8. Beniwal, G., & Singhrova, A. 2022. A systematic literature review on IoT gateways. *Journal of King Saud University-Computer and Information Sciences* 34(10): 9541–9563.

9. Ejaz, W., Naeem, M., Shahid, A., Anpalagan, A., & Jo, M. 2017. Efficient energy management for the internet of things in smart cities. *IEEE Communications Magazine* 55(1): 84–91.
10. Cui, L., Xie, G., Qu, Y., Gao, L., & Yang, Y. 2018. Security and privacy in smart cities: Challenges and opportunities. *IEEE Access* 6: 46134–46145.
11. Saleem, S.I., Zeebaree, S., Zeebaree, D.Q., & Abdulazeez, A.M. 2020. Building smart cities applications based on IoT technologies: A review. *Technology Reports of Kansai University* 62(3): 1083–1092.
12. Shyam, G. K., Manvi, S. S., & Bharti, P. 2017. *Smart waste management using Internet-of-Things (IoT). In 2017 2nd International Conference on Computing and Communications Technologies (ICCCT),* pp. 199–203, Chennai, India.
13. Abdullah, N., Alwesabi, O.A., & Abdullah, R. 2019. IoT-based smart waste management system in a smart city. In *Recent Trends in Data Science and Soft Computing: Proceedings of the 3rd International Conference of Reliable Information and Communication Technology (IRICT 2018),* pp. 364–371, Kuala Lumpur, Malaysia.
14. Samih, H. 2019. Smart cities and internet of things. *Journal of Information Technology Case and Application Research* 21(1): 3–12.
15. Karthi, S., Narmatha, N., & Kalaiyarasi, M. 2022. Industrial IOT (IIOT) based secure smart manufacturing systems in SME'S. In *Cyber Security Applications for Industry 4*.0, Taylor & Francis. ISBN: 9781003203087.
16. Saravanan, S., Kalaiyarasi, M., Karunanithi, K., Karthi, S., Pragaspathy, S., & Kadali, K.S. 2021. IoT based healthcare system for patient monitoring. In *Lecture Notes in Networks and Systems*, Springer. ISSN: 23673389 23673370, pp. 445–453.
17. Poongodi, T., Balamurugan, B., Sanjeevikumar, P., & Holm-Nielsen, J. B. 2019. Internet of things (IoT) and E-healthcare system: A short review on challenges. *IEEE India Info* 14(2): 143–147.

7 Security Challenges for Artificial Intelligence and IoT in Adoptability of Smart City

Archana Sharma, Purnima Gupta, and Aswani Kumar Singh

7.1 INTRODUCTION

As a new paradigm brought about by recent advancements in communication technologies, Smart cities have emerged to dynamically manage municipal resources and improve citizen amenities and standard of living. Smart cities require a variety of elements, including persistent smart sensors, broadband connections, database technology, and sophisticated cloud services, to gather, transmit, preserve, and intelligently analyse real-time information. New tools and services for improving residents' daily lives in the areas of decision-making, energy use, transportation, healthcare, and education are possible in smart cities. Regardless of the potential of smart cities, concerns about security and confidentiality need to be carefully addressed. Smart city is one of the major deployment areas of IoT. The usage of smartphones, RFIDs, actuators, and sensors, apart from many other smart city technology interfaces, endorses cloud as well as IoT-associated services. The universal standard technique consequently has to be filled by a bridge that connects them. In spite of this, adopting cutting-edge technology and solutions, alike artificial intelligence, will enable us to significantly increase the usability of our digital platforms and tools as a front-line decision-support system for the urban environment. The cities stand to gain from the implementation of AI because of the resources created as part of the Smart Cities Mission and the intricate dynamics of the urban ecology. This study explores the AI as well as IoT technologies importance, these two technologies have a considerable impact on how smart applications are developed and deployed. Due to its varied requirements, there are still a lot of challenges with regard to its deployment. The authors of this study attempted to combine IoT with artificial intelligence for smart city expansion. They also presented the different categories of security threats and their mitigation strategies in reference to smart cities.

DOI: 10.1201/9781003459835-7

7.1.1 Artificial Intelligence

Regardless of the potential of smart cities, concerns about security and confidentiality need to be carefully addressed. Often, machine learning and other aspects of technology are included under the category of artificial intelligence. Hardware and software are required for machine learning algorithms in association with AI. By Using AI features, generative text, audio, and various other media can all be developed.

7.1.2 AI Approaches in Smart City

In smart cities, as depicted in Table 7.1, there are several categories of AI technologies. A subcategory of AI called machine learning enables computer programs to improve their prediction skills without being expressly designed for that purpose. Taking input as historic data, machine learning algorithms predict innovative output standards. NLP, as a subcategory of artificial intelligence, studies computer-human communication. The use of AI is to produce results in the interdisciplinary field of computational linguistics and computer science known as speech recognition. Some of the alternative titles are voice recognition, speech to text. The fields of NLP and automated voice recognition fall under the supervision of artificial intelligence. With the support of artificial intelligence techniques an expert system solves various problems in a precise domain where human skill is normally needed. Although the goal of robotics is to create machines without human support where artificial intelligence (AI) is the procedure of designing arrangements that represent the learning processes of the human mind and support in the decision-making.

7.1.2.1 Machine Learning (ML)

When robots employ machine learning (ML) techniques, they can use mathematical concepts to examine data and artificially inflate its performance in a specific task without requiring explicit training (Alpaydin, 2016; Ray, 2019). Using input data, this

TABLE 7.1
AI Approaches in Smart City

Machine learning types	• Deep learning • Predictive analytics
Natural language processing	• Transformation • Data mining • Classification besides clustering
Verbal communication	• Verbal to transcript • Transcript to verbal
Visualization	• Image acknowledgement • Machine identification
Robotics	
Expert system	

kind of AI enables robots to recognize patterns as well as build estimates. Each of the supervised, and unsupervised, reinforcement, along with deep learning machine learning algorithms, has advantages as well as disadvantages. They have numerous applications, including speech and image recognition, predictive analytics, and other areas (Tyagi & Chahal, 2022). Data Analysis using ML algorithms in smart cities can be done to make estimates built on patterns as well as leanings that people can notice (Ullah et al., 2020).

7.1.2.2 Natural Language Processing

The objective of NLP as an area AI has been used to expand man-machine interaction besides communication with the use of natural language (Tyagi & Bhushan, 2023; Chowdhary, 2020). In the computer science field it is conceivable for computers to understand and develop the vocal and written human language. The clustering of documents, language conversion, and analysis of sentiments are the few tasks that can be done by NLP systems, besides that it can handle the massive amounts of transcriptions, data of language processing as well as emojis (Chowdhary, 2020). Natural Language Processing may be utilized in a variety of ways in smart cities like to influence customer support, voice agents, chatbots, and social media analysis with the support of text-based resources for the identification of expected safety risks, analysis of various urban planning projects by respondent's responses, multilingual assistance, and social media data examination for the identification of traffic problems and provide updated information to drivers(De Oliveira et al., 2020; Wang et al., 2021). The employment of NLP in smart city consistently improves the productivity and accessibility for the furtherance of necessities of the citizens (Sánchez-Ávila et al., 2020; Parkavi et al., 2021).

7.1.2.3 Computer Vision

CV algorithm is a sequence of exploratory and statistical methods. Artificial intelligence utilize CV algorithms to permit robots to analyse and for understanding of visual data, collected from their environment (Szeliski, 2022; Kothadiya et al., 2021; Davies, 2017). These algorithms can classify relevant characteristics in pictures as well as other visual data. CV is a critical field of research in AI. Other CV processes comprise segmentation procedures to separate an image into specific objects or parts and feature recognition algorithms to identify specific visual elements in a picture. By analysing security camera feeds to identify risks to public safety, streamlining waste collection routes, cutting down on resource usage, and more, smart cities may become safer as well as added efficient., and sustainable by using CV algorithms to automate the processing of visual data from multiple sources (Feng et al., 2019; Forsyth, 2002; Bhatt et al., 2021).

7.1.2.4 Robotics

One of the first domestic robots to enter our homes was the iRobot Roomba (Forlizzi & DiSalvo, 2006), which debuted there about two decades ago. Robotic floor cleaners, lawnmowers, and vacuum cleaners operated by themselves are now commonplace in homes. According to Wirtz et al. (2018), these machines have entered our daily lives as well as current industrial advancements, for instance, better vision mechanisms, added precise localization, optimized energy utilization, and enhanced machine learning. Nowadays, the practice of robotic systems has increased, moving from the private, enclosed space of a house to the complex, public area of a metropolis. Numerous dissimilar robots are currently present in our cities and are working. Automation in logistics is quite popular, both for moving people and commodities (Salonen & Haavisto, 2019; Schneider, 2019). There are automated driving delivering robots operating in many cities and on college campuses (Hoffmann & Prause, 2018), and Companies are pushing to include flying robots in expedited delivery. Trials of self-driving pods for personalized public transit have been conducted in numerous places worldwide (Ulmer & Thomas, 2018). Cities have also adopted autonomous security robots (Loke, 2018). They perform the same role as security cameras and additionally provide a user interface for citizens contacting emergency services (Lopez et al., 2017). Robots can effectively complement public authorities' operations in handling and containing the iRobot. Roomba was among the first native robots to infiltrate our homes (Forlizzi & DiSalvo, 2006), making its debut there around 20 years ago. Robotic as the ongoing healthcare crisis has shown, robots can effectively assist public authorities' actions in dealing with and covering crisis. Robots that are accustomed to sanitising large public spaces and inform individuals about their healthcare obligations are two examples (Guettari et al., 2020; Sathyamoorthy et al., 2020).

7.1.2.5 Expert System

Expert systems are computer programs designed to tackle complex hitches and offer decision-making powers comparable to human experts. By using reasoning and inference techniques, the system can retrieve information from its knowledge base in response to user requests. A subset of artificial intelligence, the first-ever expert system (ES) was created in 1970 and was the first use of AI that worked. It can solve even the most challenging problems, like an expert, by using the knowledge stored in its knowledge base. The system uses both statistics and assumptions to assist in decision-making for complicated problems, just like a human expert does. These provisions are intended for a certain industry, such as the medical or scientific fields. Some instances of current expert systems in use are as follows: Early cancer detection is possible with the Cadet (Cancer Decision Support Tool). DENDRAL helps chemists identify unknown organic compounds. Dxplain, a clinical assistance system, is used to diagnose different conditions.

7.2 AI IMPACT IN SMART CITIES

As per the Kar et al. (2019), technology is a vital component of smart city, and creative technological approaches successfully help cities grow smarter. ICT is utilized in "smart cities" to automate procedures and improve living conditions for urban dwellers. Moreover, it makes practice of integrated intelligence technology to advance responsive governance and enhance municipal infrastructure, allowing individuals to participate in the management of their cities. Thanks to a number of contemporary skills and methodologies, smart service standards can improve operations as well as efficiency in such kinds of industries which include healthcare, energy, transport and education (Herath et al., 2021). While traditionally, the goal of smart city systems along with technology has been to generate data and discuss fresh insights into the complexity and dynamics of a city, and cities may use AI to build improved decisions by using data and expertise, which elevates cities to a new level. Metropolitan artificial intelligence (AI) may be described as "artefacts functional in cities that are proficient in gathering and interpreting data about the nearby urban environment and, in the end, applying the knowledge gained to behave sensibly in line with predetermined goals in challenging urban scenarios where certain data may be lacking or insufficient". Smart city AI applications can be categorized using the following seven criteria:

- AI for governance, including disaster management, individualized subsidy provision, and urban planning.
- Artificial intelligence (AI) for safety, security, and healthcare, including smart police, individualized healthcare, reducing noise and annoyances, and improving cybersecurity.
- AI improves public appointment besides informative decision-making with the utilization of locally accurate, validated data.
- Artificial intelligence for the economy, including improved resource efficiency, collective services, well-organized supply chains, and personalized solutions.
- Artificial intelligence in logistics and transportation, including self-sufficient and eco-friendly vehicles, intelligent parking and routing, supply chain resilience, and traffic management.
- Artificial intelligence for infrastructure, such as improved organization distribution, usage, and preservation, together with waste as well as water management, electrical networks, transportation, and urban lighting.
- Artificial intelligence survives environmental factors like air quality, urban farming, and biodiversity preservation.

AI contribution in various smart city domains, approved by Bellini et al. (2022) and Herath et al. (2021). Table 7.2 and Figure 7.1 cover the AI applications in the various smart city domains.

According to a study conducted in 2018 by Ingwersen and Serrano-López, artificial intelligence has been utilized in smart city investigation since 2008. Sadly, less developed nations are pushing the UN's supportable expansion ambitions using AI, which has linked it to global sustainable advancements. (SDG). The artificial intelligence practice in smart city offers several benefits, such as improved energy and waste management, less traffic jams, and lower levels of noise and pollution. The primary objectives of smart city initiatives and technology have been data creation and a knowledge of the dynamics and complexity of cities. AI scale cities toward a new level by enabling them to use such knowledge and insights to support decision-making. AI has the potential to improve the social welfare, resilience, sustainability, and vibrancy of urban life. Because AI-built smart city programmes are spreading

TABLE 7.2
AI Applications in Smart City Domain

	Domain of Smart City Domain	Explanation	Involvement
Smart	Education	A learning paradigm known as "smart education" is geared toward meeting the needs of future generations of digital natives	Simulated reality instructional platforms, smart classes, student monitoring systems, special education needs students' Smart Library resources
	Mobility	An automated logistics and transportation network is known as a smart mobility network	Traffic control, autonomous and sustainable transportation, resilience of the supply chain, parking, and route planning
	Healthcare	Wearable technology, IoT, and mobile internet are all used in the healthcare delivery system known as "smart healthcare" to continually access data as well as link society, assets, and organizations	Patient monitoring, smart health monitoring tracking, telemedicine, telemedicine, and smart clinic E-health data and predictions of pandemics
	Surroundings	The notion of creating a room with integrated screens, processing power, and sensors to enable individuals to better see and manage their surroundings	Weather tracking, waste disposal, flood control, smart irrigation, solar cells, and environmental monitoring
	Governance	Technology utilization and innovation in government institutions to enhance planning and decision-making are known as "smart governance"	E-governance, disaster prevention, policies for making decisions, management, city planning
	Infrastructures and services	"Smart living" pertains to using city infrastructure to enhance people's quality of life	Intelligent residences and structures. Three things that are smart: education, tourism, and police
	Financial prudence	"Smart economies" are defined as those in which technology advancement, sustainability, social wellbeing, and economization are highly valued	Smart business, online shopping, and retail smart purchasing, an online marketplace for peers, and mutually beneficial labour services intelligent supply chains, shrewd sharing solutions

quickly, researchers, policymakers, and practitioners are always searching for fresh data and methods for building smart cities. Artificial intelligence will have a significant influence on society, much like many previous revolutionary inventions in human history. In many urban policy circles, especially among urban administrators and planners searching for technological answers to pressing issues of urban sprawl, AI has grown in popularity. Furthermore, despite the advantages of automation and efficiency that smart city AI systems bring, they also present moral, legal, and regulatory issues, including discrimination in service provision. It is critical to assess

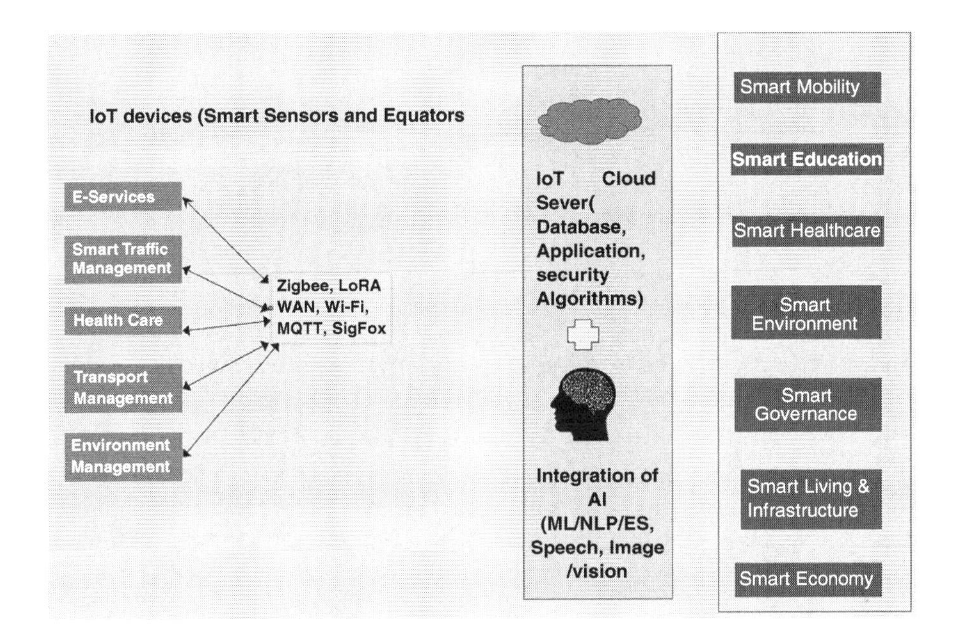

FIGURE 7.1 AI applications in Smart City domain.

the AI algorithms benefits, drawbacks, and consequences on smart cities due to the divergent opinions expressed by many specialists.

7.3 COMPARATIVE ANALYSIS OF IOT AS WELL AS AI APPLICATIONS IN SMART CITIES

To properly understand the benefits and drawbacks of IoT as well as AI deployments in smart cities, a comparative study is required. This section offers a comparative analysis of both technologies, comparing their ideas, methodology, operational procedures, impacts, security issues, and attack vectors. Table 7.3 presents a comparative study of IoT as well as Artificial Intelligence with consideration of various parameters in smart cities for implementation.

Concepts: As a wide network of interconnected devices in smart cities, Internet of Things are often functioning with sensors and actuators. These devices exchange data and communicate with one another to facilitate data-driven decision-making and enhance urban services (Angelakis et al., 2016; Zanella et al., 2014). In contrast, artificial intelligence (AI) in smart cities look for to replicate human intelligence by permitting machines for the identification of patterns, gather information, and finally convert into a logical decision (Batty et al., 2012; Rouse, 2019).

a. **Approaches**: Data assortment and communication through a network of interconnected devices is prime objective of Internet of Things. Real-time data has been collected by IoT devices about the environment,

TABLE 7.3
Comparative Study of Key Aspects Using IoT as well as AI in Smart Cities

Topic	IoT	AI
Concepts	Network of interconnected devices	Simulation of human-like intelligence
Approaches	Data collection and communication	Data processing and learning from data
How it works	Sensors collect data, transmitted to hubs	Algorithms process data, learn patterns
Impact	Efficient resource allocation and services	Enhanced decision-making and optimization
Security challenges	Weak authentication, lack of encryption	Data privacy breaches, model poisoning
Types of attacks	DDoS, man-in-the-middle, phishing	Ransomware, AI model poisoning, data breach

energy consumption, traffic design with other tasks and advancing it to cloud computing platforms or centralized centres for additional processing and investigation (Zanella et al., 2014; Angelakis et al., 2016), smart cities using AI are primarily concerned on data processing and draw conclusions from it. AI systems search for patterns in data that has been utilized by deep learning and machine learning and t further draw decisions or predictions (Batty et al., 2012; Chui et al., 2018).

b. **How It Works**: Traffic sensors, weather monitors, security cameras, etc are IoT devices in smart cities that uninterruptedly gather data from their environments. These data are managed and used for multipurpose at cloud platforms or central nodes. AI in smart cities underlies on the concepts of judgement and learning. AI algorithms utilize knowledge to learn, anticipate, and make wise decisions from the data which has been received from IoT resources (Angelakis et al., 2016; Batty et al., 2012; Gubbi et al., 2013).
c. **Impact**: IoT has a significant impact on smart cities to improve urban services and better resource management (Zanella et al., 2014; Gubbi et al., 2013). Resource distribution, energy waste drop, traffic control reduction, and waste management process optimization would be improved with the incorporation of IoT-driven solutions in smart cities. However, because AI enhances optimization and decision-making, it has an influence on smart cities (Batty et al., 2012). Artificial intelligence (AI) technologies provide real-time data analysis, predictive analytics, and automation, hence enhancing decision-making in many domains (Hashem et al., 2016).
d. **Security Challenges**: Smart city applications of IoT and AI both confront unique security challenges. IoT is susceptible to physical manipulation, firmware vulnerabilities, inadequate authentication, and unencrypted data transport (Roman et al., 2011; Sivanathan et al., 2018). Contrarily, AI struggles with issues including data privacy violations, adversarial attacks like model poisoning, bias in AI models, and the inexplicability of AI (Adeli & Jiang, 2008; Burrell, 2016).

Types of Attacks: Intelligent Cities AI and IoT are open to several types of attacks. Phishing, man-in-the-middle, and DDoS attacks are all possible on IoT devices, and they can disrupt services, intercept data, and result in financial loss (Sivanathan et al., 2018). Attacks such as ransomware, AI model poisoning, and data privacy violations affect data encryption, decision-making, and public trust in AI implementations (Roff & Moyes, 2016).

IoT and AI both play crucial roles in the creation of smart cities, as the comparative investigation has shown in Table 7.3. AI adds intelligence to the obtained data, whereas IoT focuses on data gathering and transmission, enabling informed decision-making. AI and IoT, both have security issues and are prone to various cyberattacks. Being aware of the benefits and drawbacks of these technologies, smart cities will be able to optimize the IoT and AI together to build more efficient, safe, and sustainable urban environments.

7.4 SECURITY CHALLENGES IN IOT AND AI IMPLEMENTATIONS

The impact of considerable research on IoT and AI has been made for the employability of their applications in smart city development, especially the security risks. Machine learning and deep learning may be applied in artificial intelligence (AI) to examine the massive amount of data produced due to urban environments to build intelligent decision (Batty et al., 2012; Adeli & Jiang, 2008). The incorporation of these features may improve the sustainability, productivity and liveability in smart cities. Waste management, public security, traffic control, health, etc., number of applications which have been affected by AI (Zanella et al., 2014; Bibri & Krogstie, 2017; Mohanty et al., 2016). As a backbone of data collection for smart cities, Internet of Things (IoT) devices like sensors and actuators provide a present vision of various elements of the urban environment (Zanella et al., 2014; Atzori et al., 2010).

Various studies focus on the impact of IoT incorporation in the development of infrastructure and public services (Bibri & Krogstie, 2017; Rathore et al., 2016; Al Nuaimi et al., 2015). Security may be a major concern during IoT and AI implementation in collaboration in smart cities **as** IoT devices are connected over public networks whereas AI systems contain sensitive data, which may be inclined to security issues. IoT devices which have been employed in smart cities have numerous security threats, like encryption issues, firmware flaws, authenticity problems, device upgradation etc. (Roman et al., 2011; Sivanathan et al., 2018). Even AI-driven systems have similar security issues like model exterminating and data breach. Similar issues have been raised regarding security issues with AI-driven systems, such as data privacy violations, model poisoning, and AI models biases (Adeli & Jiang, 2018). Numerous cyberattacks have also been looked into as potential threats to smart cities, including man-in-the-middle assaults, phishing, DDoS, ransomware, and AI model poisoning. These attacks could compromise citizen privacy, significantly hinder the operation of smart cities, and have an impact on making choices. Robust authentication and encryption, regular firmware upgrades, network segmentation, safe AI model training, AI bias reduction, and AI explainability are just a few of the remedies that have been proposed to address these worries (Dhanvijay & Patil, 2018). Previous Studies and research works indicated that IoT could be optimally utilized in smart cities, whereas AI will allow and support sustainable urban development by implementing strong security measures in place and fostering a cybersecurity aware culture (Da Xu et al., 2014). Future research will anticipate concentrating on improving security procedures, creating standardized frameworks for IoT and AI, and discovering novel uses for these technologies if smart cities are being formed.

The massive data produced by IoT should be processed and stored securely to prevent privacy breach of citizens. Similar data privacy issues have been stated by other researchers, who have advised the need for robust encryption, secure communication, and appropriate access control methods to protect sensitive data. Researchers have found many safety and privacy concerns in IoT devices that could be misused by maliciously intended parties. AI systems are also vulnerable, and AI decision-making processes or compromised system operations can be compromised by these vulnerabilities (Adeli & Jiang, 2008; Roff & Moyes, 2016). There is one common finding amongst these researches is that IoT devices are very vulnerable due to their easily guessable default user credentials. Any hacker can breach the security or authentication mechanism of IoT devices therefore there is a requirement for a strong security mechanism.

Few studies have explained data manipulation during data transfer across networks between IoT devices. One of the primary causes of this danger is the missing data encryption used while transferring, which exposes the data to unauthorized access and unauthorized data manipulation. The routine firmware upgrade may improve the IoT device's security (Wurm et al., 2016). Frequent upgrades of firmware ensure protection from the latest security challenges. Infrequent firmware upgrades can leave an associated Internet of Things device open to security flaws. Physical security is also a major concern for IoT devices. Any malicious intender can perform physical manipulation or damage to IoT devices. So, the physical protection of IoT devices should be ensured as a priority. IoT networks are prone to network attacks like distributed denial of service (DDoS) and others. There should be a strong security mechanism so that an effective safeguard for IoT networks could be established (Lawal et al., 2021).

In the study of challenges in interoperability and standardization in Internet of Things systems it is found that issues in IoT networks are related to a vast range of IoT devices with many different protocols and standards, which is the cause of these issues. To provide seamless integration and efficient operation inside smart city ecosystems, a consistent approach to IoT device standards is critically needed. IoT network size would have been increased due to no restriction in attaching new IoT devices; therefore, a Resilient network design is required to handle the increased number of devices without losing speed or usefulness (Zyrianoff et al., 2018). It has come to knowledge that AI in smart cities is at risk because any malicious intender can manipulate or misuse the AI system. Few researches have shown that model poisoning could take place unauthorizedly by hackers or malicious intenders during AI model training highlights the importance of AI systems from such cunning attacks (Table 7.4).

TABLE 7.4
Notable Security Attacks on IoT and AI Implementations

Year	Attack Name	Description	Impact
2010	Stuxnet	A malevolent computer worm that predominantly affected Internet of Things (IoT) systems at Iran's nuclear plants by targeting supervisory control and data acquisition (SCADA) systems	Caused substantial damage to Iran's nuclear programme
2013	Target data breach	In one instance of an IoT-related security incident, hackers used credentials taken from an HVAC business to access Target's gateway server	40 million consumers' credit card information was exposed
2016	Mirai Botnet	Malware that may be used as a component of sa botnet in extensive network attacks to create remotely controlled bots out of Internet of Things devices	Caused widespread Internet outages
2016	Tesla Autopilot Hack	Researchers remotely manipulated the Tesla Model S's autopilot function, an AI-driven feature, highlighting vulnerabilities in AI systems	Demonstrated the potential danger of AI system manipulation
2018	Casino Fish Tank Hack	Hackers accessed the high-roller database of a casino via a smart fish tank, an IoT device	Breached the database and extracted data
2019	AI Deepfake Attacks	Increasing use of AI to generate deep fakes for malicious purposes, posing significant challenges to individuals and businesses	Manipulated public opinion, committed fraud, and violated privacy
2020	Ripple20	Ripple20 was a series of zero-day vulnerabilities discovered in a widely used TCP/IP software library called Track. This library was used on various IoT devices	Attackers were able to carry out denial-of-service attacks, remotely execute code, and obtain unauthorized access to impacted devices as a result of the vulnerabilities
2021	ProxyLogon	ProxyLogon was a series of critical vulnerabilities in Microsoft Exchange Server, which is widely used for email communication. While some IoT devices running on vulnerable Exchange servers could have been compromised.	Attackers exploited these vulnerabilities to gain unauthorized access to the servers, leading to data breaches and potential device control
2022	MosaicRegressor	Android smartphones are the target of the Mosaic Regressor. Usually, connections to malicious websites found in phishing messages are used to distribute it	Malware may gather a lot of information after its installed on a device, such as call records, text messages, location information, screenshots, and audio

In conclusion, the literature review confirms that IoT and AI implementations in smart cities pose significant security challenges. The existing literature provides valuable insights into these challenges and potential countermeasures.

7.4.1 Security Issues in IoT Integration with AI Applications in Smart City

Although using AI besides IoT offers numerous benefits for smart cities, there are unique security vulnerabilities. Potential weaknesses that adversaries could exploit are created by the integration of massive interconnected networks, data collection tools, and AI-driven decision-making systems. This section examines the major security risks that IoT and AI-related smart cities confront, along with possible consequences and mitigating measures.

7.4.1.1 Security Issues in Smart Cities IoT Devices

City Smart security flaws can occur in IoT devices, including environmental sensors, security cameras, and traffic sensors. IoT devices are vulnerable to unwanted access and control due to lax authentication techniques, such as default or easily guessable passwords. Sensitive data could be intercepted and altered by malevolent parties if the data transfer is not encrypted. IoT device firmware susceptibilities can be manipulated to increase unauthorized access or launch cyberattacks. These vulnerabilities are frequently caused by obsolete or unpatched firmware. IoT devices are exposed to more possible security hazards if they are not swiftly updated with the most recent security fixes. Another serious issue is physical tampering. IoT devices are vulnerable to hostile physical attacks. Who can alter their functioning, corrupt their data, or even introduce malicious code (Roman et al., 2011; Sivanathan et al., 2018). Table 7.5 lists the security flaws along with a description and possible consequence.

TABLE 7.5
Security Issues in Smart City IoT Devices

Security Issue	Description	Potential Impact
Weak authentication	Use of default or weak credentials for IoT devices	Unauthorized access and control
Lack of encryption	Absence of data encryption during communication	Data interception and tampering
Firmware vulnerabilities	Unpatched or outdated firmware in IoT devices	Exploitation of known vulnerabilities
Lack of device updates	Inability to update IoT devices with the latest patches	Prolonged exposure to security risks
Physical tampering	Physical admittance to IoT devices for malicious purposes	Device tampering and unauthorized control

TABLE 7.6
Security Issues in AI-Driven Smart City Systems

Security Issue	Description	Potential Impact
Data privacy breaches	Unauthorized access to sensitive data used in AI models	Compromised citizen privacy and trust
Adversarial attacks	Manipulation of AI models with crafted inputs	Incorrect decision-making and outcomes
Model poisoning	Tainting AI models during training with malicious data	Compromised model integrity and reliability
Bias in AI models	Inherent biases in AI models leading to unfair decisions	Discriminatory or biased outcomes
AI explainability	Transparency is lacking in AI decision-making procedures	Difficulty in understanding model behaviour

7.4.1.2 Security Issues in AI-Driven Smart City Systems

Artificial intelligence there exist security vulnerabilities that may compromise the reliability and efficacy of smart city systems. Data privacy issues may occur if unapproved parties get access to private information used to develop or test AI models. This may reduce public support for smart city efforts (Da Xu et al., 2014; Buczak & Guven, 2015). Adversarial assaults entail tricking AI models into producing the wrong results by introducing well-constructed inputs. For instance, attackers may take advantage of flaws in image recognition software to misidentify or misclassify objects. Model poisoning is a strategy in which adversaries tamper with AI models' training data to produce biased or incorrect results. Biased models may have serious repercussions, supporting unjust or discriminatory practices. Because AI supervisory processes are opaque, it can be difficult to comprehend why particular judgements were taken, which undermines accountability and trust (Adeli & Jiang, 2008; Biggio & Roli, 2018a, b; Brundage et al., 2018). Table 7.6 highlights the security issues in AI-driven smart systems and their potential impact.

7.4.1.3 Kinds of Attacks in Smart City using IoT and AI Implementations

Since smart cities rely mostly on networked IoT devices and AI-driven systems, they could be targets for various cyberattacks. Vulnerabilities in these technologies could be used by malicious actors to compromise voting systems, compromise data privacy, or attack critical infrastructure. This section looks at a number of attack types that smart cities face, along with their shortcomings, consequences, typical cases, possible solutions, impact regions, and future reach. Figure 7.2 highlights the different security breaches that occur throughout the IoT and AI implementation of smart cities.

Table 7.7 illustrates several attack vectors in smart city scenarios involving IoT and AI implementation, as well as their vulnerabilities, effects, and future potential.

a. **DDoS Attack**: Many devices attacking a single network or server is known as a denial-of-service (DDoS) attack. This attack aims to interfere with regular traffic flow by flooding a targeted server or network with spoof

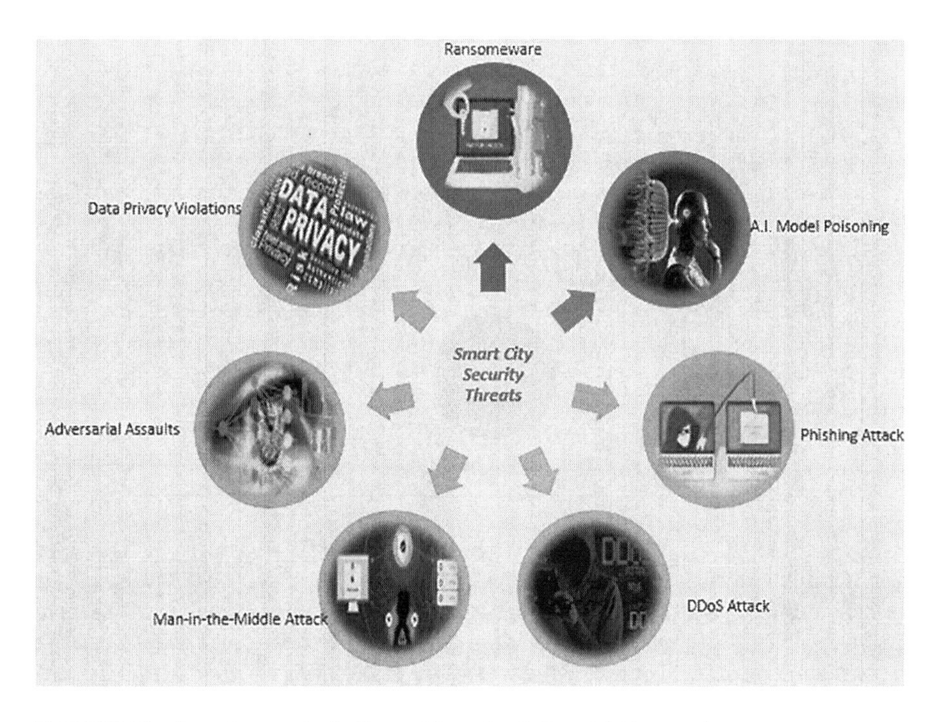

FIGURE 7.2 Security attacks in Smart City using IoT and AI integration.

requests. This puts too much strain on the network, disrupting legitimate traffics service. Malicious software compromises devices, including PCs and Internet of Things gadgets, leaving networks of Internet-connected devices open to manipulation from a distance. These attacks take advantage of these vulnerabilities. These gadgets are referred to as bots. DDoS attacks are successful because their primary attack source is botnets, or networks of compromised computers (Kolias et al., 2017).

b. **Man-in-the-Middle Attack**: It is one of the most popular types of network attacks. When an invader is able to stoop between the sender and the recipient, a communication breakdown occurs. By tricking the participants into thinking they are speaking with one another when, in reality, the attacker is in control of the communication, the attacker manipulates the communication. Frequently, there are two components to communication: the client and the server (Mehmood et al., 2017).

c. **Phishing Attack**: Phishing is the playacting to be a consistent source in online communication with the aim to obtain subtle data which includes, usernames, credit card and passwords. In General, it pretends to users that they are communicating on trusted social media websites while online payment process, or innocent people are being trick by IT administrators by asking questions frequently. Links are attached by malware-infected websites in phishing emails (Alsharnouby et al., 2015; Hadnagy & Fincher, 2015).

TABLE 7.7
Types of Attacks in Smart City IoT and AI Implementations

Attack Type	Weakness	Impact	Popular Attacks	Possible Solutions	Impact Areas	Future Scope
DDoS attack	Weak network architecture	Service disruption	Mirai Botnet attack, Dyn DNS attack	Enhanced network resilience, traffic filtering	Public services, traffic management, communication	AI-driven DDoS mitigation solutions
Man-in-the-middle attack	Weak encryption and authentication	Data interception, tampering	Eavesdropping, data alteration	Strong encryption, digital certificates, VPN	Communication, data privacy, public safety	AI for advanced intrusion detection
Phishing attack	Social engineering	Data breach, financial loss	Email spoofing, fake websites	User awareness training, multi-factor authentication	Citizen services, data privacy, financial systems	AI for real-time phishing detection
Ransomware attack	Unpatched software, poor data backup	Data encryption and extortion	WannaCry, NotPetya	Regular software updates, secure data backup	Public services, financial systems, healthcare	AI for ransomware behaviour analysis
AI model poisoning	Weak AI model validation	Misleading decisions	Misclassification, biased predictions	Robust data validation, model testing, diversity	Public safety, decision-making, critical infrastructure	AI for adversarial attacks mitigation
Data privacy breach	Inadequate data protection	Unauthorized data access	Data leakage, identity theft	Strong data encryption, access control	Healthcare, citizen information, public services	AI for anomaly-based privacy monitoring

d. **Ransomware Attack**: If certain conditions are not satisfied, hackers may breach a system or post private data on the Dark Web. Additionally, the victims become prime targets for extortion once they are forced to pay the ransom. Attacks using ransomware have the power to completely destroy a smart city's infrastructure and cause system lockdowns and slowdowns. Private information needs to be shielded from prying eyes and kept safe. This might entail data anonymization or the installation of firewalls (Radanliev et al., 2018).
e. **AI Model Poisoning**: Adversaries insert malicious or false data into the training dataset in an AI poisoning attack. The attacker makes minute adjustments that have the potential to corrupt the learning practice, which could be accomplished with the business, and as a result, it leads to inaccurate outputs or incorrect choices made by the AI model (Biggio et al., 2012; Barreno et al., 2006).
f. **Data Privacy Breach**: One of the main reasons why people value privacy is the problem of manipulative prompting and its impact on autonomy, which is the privacy risk connected with monitoring as a security measure. Smart cities are turning cities into enormous laboratories, with the main focus being on how to make people's behaviour predictable and controllable from outside. In this sense, technology can be thought of as manipulating the surroundings to learn about and impact visitor behaviour in order to improve the safety and appeal of the site (Bibri & Krogstie, 2017; Roman et al., 2013).

7.5 EXTENUATION APPROACHES IN IOT AND AI IMPLEMENTATION SECURITY CHALLENGES

For smart cities to address the security risks caused by IoT and AI implementations strong countermeasures need to be taken. This section covers the various security options and measures to defend smart city infrastructures against cyber threats and attacks. To address security concerns related to IoT and AI implementations in smart cities, a number of researchers have put forth a variety of mitigation strategies. Now let's explore these tactics., which are highlighted in Table 7.8.

7.5.1 Strong Authentication Mechanisms

Xu et al. (2019) recommended the implementation of strong authentication measures for the prevention of unauthorized admittance to IoT devices as well as schemes, such as unique, complex passwords, multi-factor authentication (MFA), and biometric authentication where appropriate.

7.5.2 Data Encryption

Zhang et al. (2020) highlighted the need for vigorous data encryption while communicating over the network and storage in order to guard data privacy.

TABLE 7.8
Mitigation Strategies for Security Challenges

Mitigation Strategy	Application	Strengths	Weaknesses	Limitations
Strong authentication mechanisms	IoT device security	Enhanced security, identity verification	Reliance on user management	User error, lost credentials
Data encryption	IoT data privacy	Data privacy, data integrity	Computational overhead	Not all data might be encryptable
Regular firmware and software updates	IoT device and AI system security	Patching vulnerabilities, updated security features	Requires regular maintenance, potential downtime	Not all devices support updates
Tamper-proof hardware and secure enclosures	Physical security of IoT devices	Physical protection, prevent tampering	Additional cost, potentially more difficult maintenance	Not all devices can be fully tamper-proof
AI for advanced intrusion detection	Network security in IoT environments	Real-time detection, pattern recognition	False positives, model training	Relies on the quality of threat intelligence
Data validation and model diversity	Security of AI models	Mitigate model poisoning, robust models	Time-consuming, expertise required	Assumes the presence of diverse, high-quality data
Auditing of AI models for bias	AI fairness	Fair decision-making, trust-building	Requires expertise, may not catch all biases	Subject to human interpretation and biases
Explainable AI techniques	AI explainability	Transparency, trust-building, user understanding	Can reduce model performance	Not all AI models are explainable

7.5.3 Regular Firmware and Software Updates

For IoT devices, Dhanvijay and Patil (2018) addressed the significance of periodic firmware updates. In a similar fashion, the software that powers AI systems needs updates in order to fix security vulnerabilities.

7.5.4 Tamper-Proof Hardware and Secure Enclosures

Strohmeier et al. (2015) highlighted the need for tamper-proof hardware and secure enclosures to prevent physical tampering of IoT devices.

Application: physical security of IoT devices.
Future Scope: Development of more resilient and tamper-evident materials and designs.

7.5.5 AI for Advanced Intrusion Detection

Roff and Moyes (2017) suggested using AI for advanced intrusion detection. AI can analyse network traffic, monitor system behaviour, and detect anomalies indicative of attacks.

Application: Network security in IoT environments.
Future Scope: AI-powered predictive threat intelligence

7.5.6 Data Validation and Model Diversity

Biggio and Roli (2018a, b) stressed the need for robust data validation and model diversity to prevent AI model poisoning.

Application: security of AI models.
Future Scope: Use of AI to detect anomalous training data

7.5.7 Auditing of AI Models for Bias

Zliobaite and Custers (2016) suggested regular auditing of AI models for bias to ensure fairness in AI decision-making.

Application: AI fairness.
Future Scope: Development of AI auditing tools for non-technical stakeholders.

7.5.8 Explainable AI Techniques

In order to improve transparency in AI decision-making, Doshi-Velez and Kim (2017) emphasized the significance of explainable AI methodologies.

Application: AI explainability.
Future Scope: Improvement of techniques for interpreting complex AI models.

The mitigation strategies for security challenges in smart cities employing IoT and AI span several domains. Strong authentication mechanisms for IoT device security bolster security by verifying identities but can be undermined by user management issues, such as lost credentials. Data encryption ensures IoT data privacy and integrity, although computational overheads may limit its applicability, and not all data might be suitable for encryption. Regular firmware and software updates patch vulnerabilities and provide updated security features, but this requires maintenance and could potentially lead to downtime, with some devices not supporting updates. The physical security of IoT devices is enhanced through tamper-proof hardware and secure enclosures, though additional costs and maintenance difficulties may arise, and not all devices can be entirely tamper-proof. Artificial Intelligence (AI) is used in network security for enhanced intrusion detection. It provides real-time detection and pattern identification, but it may also produce false positives, and its effectiveness depends on the quality of threat intelligence. Data validation and model diversity strengthen the security of AI models and mitigate model poisoning, yet they are time-consuming and require expertise. Regular auditing of AI models for bias ensures fairness and builds trust, but it requires expertise and may not catch all biases, making it subject to human interpretation and biases. Finally, explainable AI techniques offer transparency and promote trust, but they can potentially reduce model performance, and not all AI models are explainable (Wurm et al., 2016; Buczak & Guven, 2015; Adeli & Jiang, 2008).

7.6 CONCLUSION

Cities have changed significantly as a result of the creation and employment of several notions, which include inclusive, sustainable, and resilient cities. AI and the Internet of Things are two vital technologies that might convert cities into sustainable smart cities. The advancement of ML techniques is unquestionably essential to the development of AI. It will clear the way for handling real-time, high-frequency applications in smart cities. However, it is challenging to understand how such approaches function when underlined two important and challenging subjects that were examined in this study were smart cities "privacy and cybersecurity". Cybersecurity in smart cities is still in its early stages; thus, there are still a lot of rules, strategies, structures, and technological solutions in this important domain. By integrating AI and IoT into smart cities, the research has offered a comprehensive examination of the applications, specifications, issues, and difficulties pertaining to security in those settings. Potential remedies for resolving security issues of IoT integration with AI adoption in smart city were evaluated and further examined in depth in the second study phase.

REFERENCES

Abomhara, M., & Køien, G. M. (2015). Cyber security and the internet of things: Vulnerabilities, threats, intruders and attacks. *Journal of Cyber Security*, 4(1), 65–88.

Adeli, H., & Jiang, X. (2008). *Intelligent Infrastructure: Neural Networks, Wavelets, and Chaos Theory for Intelligent Transportation Systems and Smart Structures*. CRC Press. USA.

AlNuaimi, E., Al Neyadi, H., Mohamed, N., & Al-Jaroodi, J. (2015). Applications of big data to smart cities. *Journal of Internet Services and Applications*, 6(1), 1–15.

Alpaydin, E. (2016). *Machine Learning: The New AI*. MIT Press: Cambridge, MA.

Alsharnouby, M., Alaca, F., & Chiasson, S. (2015). Why phishing still works: User strategies for combating phishing attacks. *International Journal of Human-Computer Studies*, 82, 69–82.

Angelakis, V., Tragos, E., Pöhls, H. C., Kapovits, A., & Bassi, A. (Eds.). (2016). *Designing, Developing, and Facilitating Smart Cities: Urban Design to IoT Solutions*. Springer. London.

Atzori, L., Iera, A., & Morabito, G. (2010). The internet of things: A survey. *Computer Networks*, 54(15), 2787–2805.

Barreno, M., Nelson, B., Sears, R., Joseph, A. D., & Tygar, J. D. (2006). Can machine learning be secure? In *Proceedings of the 2006 ACM Symposium on Information, Computer and Communications Security* (pp. 16–25).

Batty, M., Axhausen, K. W., Giannotti, F., Pozdnoukhov, A., Bazzani, A., Wachowicz, M., ... & Portugali, Y. (2012). Smart cities of the future. *The European Physical Journal Special Topics*, 214, 481–518.

Bellini, P., Nesi, P., & Pantaleo, G. (2022). IoT-enabled smart cities: A review of concepts, frameworks and key technologies. *Applied Sciences*, 12 (3), 1607.

Bhatt, D., Patel, C., Talsania, H., Patel, J., Vaghela, R., Pandya, S., Modi, K., & Ghayvat, H. (2021). CNN variants for computer vision: History, architecture, application, challenges and future scope. *Electronics*, 10, 2470.

Bibri, S. E., & Krogstie, J. (2017). Smart sustainable cities of the future: An extensive interdisciplinary literature review. *Sustainable Cities and Society*, 31, 183–212.

Biggio, B., Nelson, B., & Laskov, P. (2012). Poisoning attacks against support vector machines. arXiv preprint arXiv:1206.6389.

Biggio, B., & Roli, F. (2018a). Wild patterns: Ten years after the rise of adversarial machine learning. *Pattern Recognition*, 84, 317–331.

Biggio, B., & Roli, F. (2018b). Wild patterns: Ten years after the rise of adversarial machine learning. In *Proceedings of the 2018 ACM SIGSAC Conference on Computer and Communications Security* (pp. 2154–2156).

Brundage, M., Avin, S., Clark, J., Toner, H., Eckersley, P., Garfinkel, B., ... & Amodei, D. (2018). The malicious use of artificial intelligence: Forecasting, prevention, and mitigation. arXiv preprint arXiv:1802.07228.

Buczak, A. L., & Guven, E. (2015). A survey of data mining and machine learning methods for cyber security intrusion detection. *IEEE Communications Surveys & Tutorials*, 18(2), 1153–1176.

Burrell, J. (2016). How the machine 'thinks': Understanding opacity in machine learning algorithms. *Big Data & Society*, 3(1), 2053951715622512.

Chowdhary, K. (2020). Natural language processing. In *Fundamentals of Artificial Intelligence* (pp. 603–649). Springer: New Delhi, India.

Chui, M., Manyika, J., Miremadi, M., Henke, N., Chung, R., Nel, P., & Malhotra, S. (2018). *Notes from the AI Frontier: Applications and Value of Deep Learning*. McKinsey Global Institute Discussion Paper.

DaXu, L., He, W., & Li, S. (2014). Internet of things in industries: A survey. *IEEE Transactions on Industrial Informatics*, 10(4), 2233–2243.

Davies, E. R. (2017). *Computer Vision: Principles, Algorithms, Applications, Learning*. Academic Press: Cambridge, MA.

DeOliveira, N. R., Reis, H. L., Fernandes, N. C., Bastos, A. C., Medeiros, S. D., & Mattos, M. D. (2020). Natural language processing characterization of recurring calls in public security services. In *Proceedings of the 2020 International Conference on Computing, Networking and Communications (ICNC)*, Big Island, HI, 17–20 February (pp. 1009–1013).

Dhanvijay, M. M., & Patil, S. C. (2018). Internet of Things: A survey of enabling technologies in healthcare and its applications. *Computer Networks*, 144, 288–306.

Doshi-Velez, F., & Kim, B. (2017). Towards a rigorous science of interpretable machine learning. arXiv preprint arXiv:1702.08608.

Feng, X., Jiang, Y., Yang, X., Du, M., & Li, X., (2019). Computer vision algorithms and hardware implementations: A survey. *Integration*, 69, 309–320.

Forlizzi, J., and DiSalvo, C. (2006). Service robots in the domestic environment: A study of the roomba vacuum in the home. In *Proceedings of the 1st ACM SIGCHI/SIGART Conference on Human-Robot Interaction* (pp. 258–265).

Forsyth, D. A. (2002). *Computer Vision: A Modern Approach.* Prentice Hall Professional Technical Reference: Hoboken, NJ.

Gubbi, J., Buyya, R., Marusic, S., & Palaniswami, M. (2013). Internet of Things (IoT): A vision, architectural elements, and future directions. *Future Generation Computer Systems*, 29(7), 1645–1660.

Guettari, M., Gharbi, I., and Hamza, S. (2020). Uvc disinfection robot. *Environmental Science and Pollution Research International*, 28, 40394. doi:10.1007/s11356-020-11184-2.

Hadnagy, C., & Fincher, M. (2015). *Phishing Dark Waters: The Offensive and Defensive Sides of Malicious Emails*. John Wiley & Sons.

Hashem, I. A. T., Chang, V., Anuar, N. B., Adewole, K., Yaqoob, I., Gani, A., … & Chiroma, H. (2016). The role of big data in smart city. *International Journal of Information Management*, 36(5), 748–758.

Herath, H. M. K. K. M. B., Karunasena, G. M. K. B., Madhusanka, B. G. D. A., & Priyankara, H. D. N. S. (2021). Internet of medical things (IoMT) enabled TeleCOVID system for diagnosis of COVID-19 patients. In *Sustainability Measures for COVID-19 Pandemic* (pp. 253–274). Singapore: Springer.

Hoffmann, T., & Prause, G. (2018). On the regulatory framework for last-mile delivery robots. *Machines*, 6, 33. doi:10.3390/machines6030033. https://msrc.microsoft.com/blog/2021/03/multiple-security-updates-released-for-exchange-server/.

Kar, A. K., Ilavarasan, V., Gupta, M. P., Janssen, M., & Kothari, R. (2019). Moving beyond smart cities: Digital nations for social innovation & sustainability. *Information Systems Frontiers*, 21 (3), 495–501.

Kolias, C., Kambourakis, G., Stavrou, A., & Voas, J. (2017). DDoS in the IoT: Mirai and other botnets. *Computer*, 50(7), 80–84.

Kothadiya, D., Chaudhari, A., Macwan, R., Patel, K., & Bhatt, C. (2021). The convergence of deep learning and computer vision: Smart city applications and research challenges. In *Proceedings of the 3rd International Conference on Integrated Intelligent Computing Communication & Security (ICIIC 2021)*, Bangalore, India, 4–5 June (pp. 14–22).

Kumari, P., & Jain, A. K. (2023). A comprehensive study of DDoS attacks over IoT network and their countermeasures. *Computers & Security*, 103096.

Lawal, M. A., Shaikh, R. A., & Hassan, S. R. (2021). A DDoS attack mitigation framework for IoT networks using fog computing. *Procedia Computer Science*, 182, 13–20.

Loke, S. W. (2018). Are we ready for the internet of robotic things in public spaces. In *Proceedings of the 2018 ACM International Joint Conference and 2018 International Symposium on Pervasive and Ubiquitous Computing and Wearable Computers, UBIComp '18: The 2018 ACM International Joint Conference on Pervasive and Ubiquitous Computing*, Singapore, October 8–12, 2018 (pp. 891–900).

Lopez, A., Paredes, R., Quiroz, D., Trovato, G., and Cuellar, F. (2017). Robotman: A security robot for human-robot interaction. In *2017 18th International Conference on Advanced Robotics (ICAR), Hong Kong,* China, 10–12 July 2017 (pp. 7–12). IEEE. doi:10.1109/icar.2017.8023489.

Mehmood, Y., Ahmad, F., Yaqoob, I., Adnane, A., Imran, M., & Guizani, S. (2017). Internet-of-things-based smart cities: Recent advances and challenges. *IEEE Communications Magazine*, 55(9), 16–24.

Mohanty, S. P., Choppali, U., & Kougianos, E. (2016). Everything you wanted to know about smart cities: The Internet of Things is the backbone. *IEEE Consumer Electronics Magazine*, 5(3), 60–70.

Mosenia, A., & Jha, N. K. (2016). A comprehensive study of security of internet-of-things. *IEEE Transactions on Emerging Topics in Computing*, 5(4), 586–602.

Parkavi, A., Sowmya, B., Jerin Francis, A., Srikanth, B., Rohan, N., & Deepak, R. (2021), "SmartEval": Evaluation system for descriptive answers in examinations using natural language processing and artificial neural networks. In *Proceedings of the 2nd International Conference on Recent Trends in Machine Learning, IoT, Smart Cities and Applications: ICMISC 2021*, Hyderabad, India, 28–29 March (pp. 557–567).

Radanliev, P., De Roure, D. C., Nicolescu, R., Huth, M., Montalvo, R. M., Cannady, S., & Burnap, P. (2018). Future developments in cyber risk assessment for the internet of things. *Computers in Industry*, 102, 14–22.

Radanliev, P., De Roure, D., Page, K., Nurse, J. R., Mantilla Montalvo, R., Santos, O., … & Burnap, P. (2020). Cyber risk at the edge: current and future trends on cyber risk analytics and artificial intelligence in the industrial internet of things and industry 4.0 supply chains. *Cybersecurity*, 3(1), 1–21.

Rathore, M. M., Ahmad, A., Paul, A., & Rho, S. (2016). Urban planning and building smart cities based on the internet of things using big data analytics. *Computer Networks*, 101, 63–80.

Ray, S. (2019). A quick review of machine learning algorithms. In *Proceedings of the International Conference on Machine Learning, Big Data, Cloud and Parallel Computing (COMITCon)*, Faridabad, India, 14–16 February 2019 (pp. 35–39).

Roman, R., Najera, P., & Lopez, J. (2011). Securing the internet of things. *Computer*, 44(9), 51–58.

Roman, R., Zhou, J., & Lopez, J. (2013). On the features and challenges of security and privacy in distributed internet of things. *Computer Networks*, 57(10), 2266–2279.

Rouse, M. (2019). What is AI (artificial intelligence)? Definition from WhatIs. com. [online] Search Enterprise AI.

Salonen, A., & Haavisto, N. (2019). Towards autonomous transportation. Passengers' experiences, perceptions and feelings in a driverless shuttle bus in Finland. *Sustainability*, 11, 588. doi:10.3390/su11030588.

Sánchez-Ávila, M., Mouriño-García, M. A., Fisteus, J. A. & Sánchez-Fernández, L. (2020). Detection of barriers to mobility in the smart city using Twitter. *IEEE Access*, 8, 168429–168438.

Sathyamoorthy, A. J., Patel, U., Savle, Y. A., Paul, M., & Manocha, D. (2020). Covid-robot: Monitoring social distancing constraints in crowded scenarios. arXiv. arXiv preprint arXiv:2008.06585.

Schneider, D. (2019). The delivery drones are coming. *IEEE Spectrum*, 57, 28–29. doi:10.1109/mspec.2020.8946304.

Sivanathan, A., Gharakheili, H. H., Loi, F., Radford, A., Wijenayake, C., Vishwanath, A., & Sivaraman, V. (2018). Classifying IoT devices in smart environments using network traffic characteristics. *IEEE Transactions on Mobile Computing*, 18(8), 1745–1759.

Strohmeier, M., Lenders, V., & Martinovic, I. (2015). Lightweight location verification in air traffic surveillance networks. In *Proceedings of the 10th ACM Symposium on Information, Computer and Communications Security* (pp. 429–440).

Szeliski, R. (2022). *Computer Vision: Algorithms and Applications*. Springer Nature: Berlin/Heidelberg, Germany.

Tyagi, N., & Bhushan, B. (2023). Demystifying the role of natural language processing (NLP) in smart city applications: Background, motivation, recent advances, and future research directions. *Wireless Personal Communications*, 130, 857–908.

Tyagi, A. K., & Chahal, P. (2022). Artificial intelligence and machine learning algorithms. In *Research Anthology on Machine Learning Techniques, Methods, and Applications* (pp. 421–446). Hershey, PA: IGI Global.

Ullah, Z., Al-Turjman, F., Mostarda, L., & Gagliardi, R. (2020). Applications of artificial intelligence and machine learning in smart cities. *Computer Communications*, 154, 313–323.

Ulmer, M. W., & Thomas, B. W. (2018). Same-day delivery with heterogeneous fleets of drones and vehicles. *Networks*, 72, 475–505. doi:10.1002/net.21855.

Wang, J., Wang, M., & Song, Y. (2021). A study on smart city research activity using bibliometric and natural language processing methods. In *Proceedings of the 2021 the 9th International Conference on Information Technology: IoT and Smart City*, Guangzhou, China, 22–25 (pp. 346–352).

Wirtz, J., Patterson, P. G., Kunz, W. H., Gruber, T., Lu, V. N., Paluch, S., & Martins, A. (2018). Brave new world: Service robots in the frontline. *Journal of Service Management*, 29, 907–931. doi:10.1108/josm-04–2018–0119.

Wurm, J., Hoang, K., Arias, O., Sadeghi, A. R., & Jin, Y. (2016). Security analysis on consumer and industrial IoT devices. In *2016 21st Asia and South Pacific Design Automation Conference (ASP-DAC)* (pp. 519–524). IEEE.

Xu, L., Chen, L., & Ren, Y. (2019). Ensuring consumer privacy in the cloud. *IEEE Internet of Things Journal*, 6(2), 3243–3255.

Zanella, A., Bui, N., Castellani, A., Vangelista, L., & Zorzi, M. (2014). Internet of Things for smart cities. *IEEE Internet of Things Journal*, 1(1), 22–32.

Zhang, R., & Cai, K. (2018). Supervisor localization of discrete-event systems with infinite behavior. *IFAC-PapersOnLine*, 51(7), 361–366.

Zhang, Y., Chen, X., & Chen, Y. (2020). Ensuring data privacy in smart city based on federated learning. *IEEE Transactions on Information Forensics and Security*, 16, 2860–2870.

Zliobaite, I., & Custers, B. (2016). Using sensitive personal data may be necessary for avoiding discrimination in data-driven decision models. *Artificial Intelligence and Law*, 24(2), 183–201.

Zyrianoff, I., Borelli, F., Biondi, G., Heideker, A., & Kamienski, C. (2018). Scalability of real-time iot-based applications for smart cities. In *2018 IEEE Symposium on Computers and Communications (ISCC)* (pp. 00688–00693). IEEE.

8 A Comprehensive Survey on Security Issue and Applications of Metaverse

A. Akshaya, N. Sai Arunaa Varshini, Raja Lavanya, N. R. Naveen Babu, K. Sundarakantham, and S. Mercy Shalinie

8.1 INTRODUCTION

The prefix "meta" is a term that signifies "more comprehensive" or "transcending" in Greek, and it is the origin of the word "meta" [1]. Universe, which stands for a space/time container, is shortened to "Verse" in this sentence. Here, the users can play with friends, live, and work anywhere in the metaverse, for instance.

The metaverse provides all the money. The Times Journal claims that the Metaverse is the next digital period that will have a big impact on humanity. It is wise to exercise caution when utilizing any new technology in cyberspace, and one should think about any possible security and privacy problems from the start.

This study was undertaken to know what are the potential applications and challenges of utilizing blockchain technology in the metaverse and how can it enhance privacy, security, interoperability and trust within the virtual environment [16].

For their research paper, which was released in 2021, Yang et al. did a thorough literature review on the application of blockchain technology in the forthcoming metaverse. In addition, there are a ton of applications, such as social networking tools, non-fungible tokens to create an economic ecology within the metaverse, token (NFT) marketplaces and other resources that are made available. A Comprehensive Analysis and Solutions by Smith, J., Johnson, A., and Lee, S gives us an elaborated insight about the matters on privacy and security in the metaverse universe.

A survey of the literature on the integration of blockchain from Industry 4.0 to Society 5.0 has been provided, along with forecasts for the metaverse.

8.2 DEFINING THE METAVERSE

8.2.1 Characteristics of Metaverse

A shared virtual environment known as the "metaverse" allows users to connect in highly immersive and interactive ways with other users and the digital realm.

 DOI: 10.1201/9781003459835-8

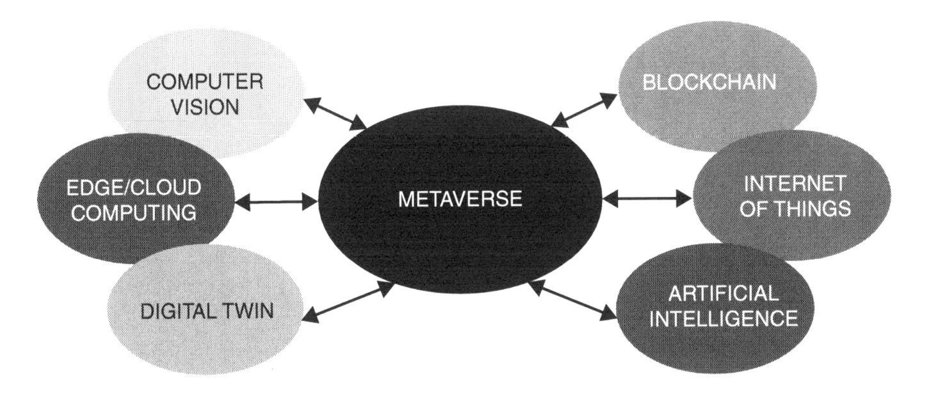

FIGURE 8.1 Framework of metaverse.

A metaverse has characteristics like:

- The metaverse is a persistent virtual world that is always accessible, regardless of the user's location or time of visit.
- Users can engage in natural and intuitive interactions with virtual things and other users in the metaverse.
- **Immersion**: Once you step into the metaverse, you'll experience an extraordinary level of immersion, thanks to advanced technologies like virtual and augmented reality that create a vivid sense of being present within the virtual world.
- **Scalability**: The complex digital worlds and millions of users are supported by the metaverse. It can grow quite quickly.
- **User-Generated Content**: The users get the ability to develop and check their own digital world [2] (Figure 8.1).

8.3 FRAMEWORK FOR THE METAVERSE

8.3.1 Edge Computing

In contrast to central data storage, edge computing involves keeping data close to the source of generation. This decreases the volume of information that must be sent over networks.

Some current edge computing technologies are as follows:

1. **Edge AI**: It indulges the application of machine learning models and algorithms to edge devices. Edge AI can increase privacy by keeping sensitive data on the device and reducing latency by processing data at the edge.
2. **Edge Storage**: As opposed to storing data in a centralized data centre, edge storage entails storing data on edge devices, such as smartphones, IoT devices, and edge servers. By keeping sensitive data on the device, this can enhance privacy, reduce latency, and data access.

3. **Edge Security**: Using intrusion detection systems and firewalls to secure local networks and edge devices is known as edge security. By thwarting dangers earlier, this can increase security.

8.3.2 Digital Twin

A digital twin depicts a solid object, a digital copy of it. There are three basic parts to it: the actual object, its digital counterpart, and the communication of data between the actual and virtual entity. Beyond our wildest dreams, this can incredibly realistically render the virtual environment. Digital twins are these three different forms of virtual replicas that combine and integrate a product's performance and creation.

Virtual models built on physics are created from all the data combined utilizing the AI algorithm [3]. Examining its virtual or digital equivalent will assist in determining and forecasting the present and future states of any physical asset.

8.3.3 Internet of Things

Through gadgets and sensors, the IoT links digital devices. For the purpose of producing an accurate representation of an object, it gathers and distributes statistics from the real world. Since the metaverse connects a variety of real-world devices through IoT, it makes it simple and seamless to adapt to the 3D environment.

The real and digital worlds becoming indistinguishable while providing a customized interface environment for IoT makes it much easier to create simulations within the metaverse, particularly with digital twins.

8.3.4 Computer Vision

We can communicate with people in the metaverse via computer vision. Extended reality products, which rely on computer vision, have the potential to create a virtual environment [7]. It uses visual data, deliberate judgements, and appropriate actions. It is an interactive setting where getting geographical data about human bodies is involved.

Technologies like digital twins, blockchain-enabled record keeping, state-of-the-art robots, and computer vision-powered images are already revolutionizing industries. Computer vision principles like gesture recognition, human pose tracking, emotion recognition, and expression analysis will assist gadgets in figuring out how people interact with their surroundings, and they will use that knowledge to create intuitive and incredibly realistic sensory experiences in the metaverse [4,5].

8.3.5 Literature Survey

Table 8.1 shows the literature survey of the various applications of blockchain technology in other fields and domains. There are various research fields that work on this disruptive technology to bring progress. Here is an overview of it (Table 8.1).

TABLE 8.1
Literature Survey

Year of Publication	Name of the Survey Paper	Summary of the Paper
2023	Blockchain integration in the Era of industrial metaverse	Improving the decentralization of authority to reduce cybercrime
2022	The blockchain-based professional certification and tourism platform with its own ecosystem in the metaverse	To build an edutainment platform for career learning
2023	A survey of blockchain and intelligent networking for the metaverse	Architecture of Metaverse and the fields of development
2022	Blockchain empowered service management for the decentralized metaverse of things	Fine data flow allocation and service selection in a decentralized network
2023	Blockchain meets metaverse and digital asset management	The metaverse can benefit from the security and privacy advantages offered by blockchain technology
2022	Enhancing wisdom manufacturing as industrial metaverse for Industry and society 5.0	Creating roadmap to blockchainized value to industrial metaverse

8.4 BLOCKCHAIN IN METAVERSE

Blockchain technology can be a way to always make transactions financially easily inside the metaverse. With the ability to purchase and sell virtual goods for real money, the metaverse allows you to interact with the entire cryptocurrency market more deeply. In the metaverse, one can always transfer and introduce one's assets and implement money.

8.4.1 Benefits of Blockchain

Another possible use of blockchain technology is to protect users from fraud and hackers of their data, implementing data security and confidentiality. Some characteristics of blockchain technology that can be used in the metaverse are as follows:

1. **Decentralization**: The decarbonization characteristics of blockchain technology ensure the creation of a reliable environment. In the future, the metaverse will become a virtual market where people will communicate, play, trade and work using various technological devices, where large amounts of heterogeneous data will be created.
2. **Smart Contracts**: It is capable of acting automatically in response to preset parameters. It can be used to control user interactions and transactions in the metaverse, including virtual item trading and completing in-game tasks. The intrinsic characteristics of the blockchain network enable smart contracts to have exceptional capabilities like automation, programmability, openness, transparency, and verifiability. This enables trusted interactions with blockchain without the requirement for a decentralized validation system [19].

3. **Tokenization**: The capacity to produce one-of-a-kind, non-fungible tokens (NFTs) on the blockchain offers ability to symbolize validations of digital properties in the metaverse.
4. Indivisible and uniqueness are NFT's most crucial characteristics, which make it ideal for representing identities that can be sold and transacted. The metaverse provides these virtual qualities with validations known as NFTs.
5. **Interoperability**: Interoperability between different metaverse systems can be facilitated by blockchain technology, which enables seamless transactions and user interactions across several virtual worlds.

The blockchain's interoperability makes it simpler for users to move between virtual worlds.

New blocks are generated and verified by a consensus process [17].

PoW consensus is the preferred consensus technique for most cryptocurrencies. PoW generation might be an erratic, unlikely procedure. Before a trustworthy proof of work can be established, a significant amount of trial and error is required. The evidence of work is based on a mathematical puzzle with an easily proven solution.

All things considered, the decentralized, secure, and programmable aspects of blockchain technology can provide a foundation for innovative and original metaverse experiences and business models.

8.5 CRYPTO'S FIT INTO METAVERSE

Because they are decentralized, cryptocurrencies have a potential to be significant players in the metaverse. Users may be able to transact through them on different virtual worlds and platforms. Cryptocurrencies may end up being a more affordable option than traditional media in a metaverse environment where transactions can be carried out without the assistance of banks or payment processors.

Metaverse cryptocurrencies typically employ blockchain technology. Two of the most well-known security features of blockchain are decentralization and immutability. Since transactions are recorded on a distributed ledger, it is difficult for hostile actors to modify data. There may still be potential for flaws like 51% attacks or smart contract exploitation because no system is 100% safe. Blockchain technology ensures transaction transparency; however, privacy concerns are often highlighted. Some users may choose to keep their financial transactions private because blockchain transactions are public information. Aiming to allay these worries are more recent blockchain systems and cryptocurrencies.

Governments may implement new rules to address security and privacy concerns as cryptocurrencies and metaverses gain popularity. How cryptocurrencies are utilized in the metaverse may change if certain rules are followed.

8.5.1 Involvement of Cryptos

Also, cryptocurrencies can be used to produce new business models for metaverse platforms, similar to through the use of decentralized independent associations. (DAOs) that are governed by smart contracts and incentivize participation through cryptocurrency prices. Overall, cryptocurrencies can provide a means of conducting flawless and secure deals in the metaverse, as well as enabling new forms of decentralized and community-driven profitable exertion [3]. All along, we've seen how blockchain and cryptocurrency contribute to metaverse. Further, let us see how blockchain. The metaverse has the implicit to significantly impact the development and relinquishment of blockchain technology in several ways.

Within virtual worlds and metaverse platforms, tokens with blockchain-based technology are used as native currencies [8]. Within the metaverse, users can earn, buy, or exchange these virtual currencies to pay for virtual goods, services, and assets. A decentralized approach to ownership within the metaverse is now possible because of cryptocurrencies and blockchain technology. True ownership of users' virtual assets is possible, and the ownership records are safely stored on the blockchain. Virtual property has grown to be a valued commodity in virtual worlds with large-scale landscapes. It is simpler to buy, sell, and transfer virtual land using cryptocurrency thanks to blockchain technology, which enables clear and secure property ownership records

8.5.1.1 Relinquishment of Cryptocurrencies

The use of cryptocurrencies and other digital means is a crucial element of numerous metaverse surroundings. As the metaverse continues to grow in fashionability, this may drive lesser relinquishment of cryptocurrencies and other blockchain grounded means.

Decentralization and Democratization: The decentralized nature of the metaverse aligns with the core values of blockchain technology, which prioritizes decentralization and democratization of digital systems. The growth of the metaverse may lead to the creation of further decentralized and community-driven blockchain systems.

Virtual Property and Asset Power: The metaverse relies heavily on the power and exchange of virtual means, similar to in-game particulars and digital real estate. The use of blockchain-grounded commemoratives to represent power of these means can give a secure and transparent means of exchange that's resistant to fraud and theft.

The metaverse is a decentralized virtual environment, and many systems are investigating the use of blockchain technology to create decentralized, stoner-possessed metaverse platforms. These platforms may allow for less stoner autonomy, seclusion, and control over virtual means and relationships. Overall, the expansion of the metaverse may contribute to the abandonment of blockchain technology and the development of new blockchain-based activities and use cases [15]. Again, the adoption of blockchain technology can assist to improve the security, transparency, and stoner experience of the metaverse.

8.6 BLOCKCHAIN INGRAINMENT IN METAVERSE

From a technical aspect, this part covers cutting-edge blockchain-based options for the metaverse, spanning data gathering, storage, sharing, interoperability, and privacy protection (Figure 8.2).

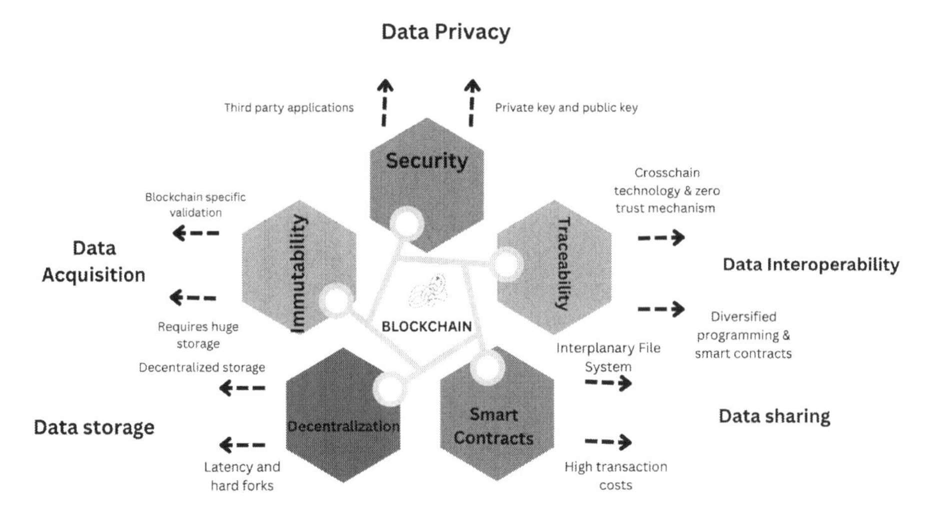

FIGURE 8.2 Technical perspectives of blockchain.

8.6.1 Technical Aspects

8.6.1.1 Data Acquisition

The data acquiring method in virtual reality is critical. Customer's sensitive details are acquired during transactions.

The data collection technique for the metaverse ecosystem is critical. When a customer completes a purchase, some of their sensitive information is collected. In the metaverse, private data such as users' biometric positions and motions must also be collected in order to build virtual identities.

Blockchain, as a distributed ledger technology, enables the collection of legitimate information and records through transactions. Every action on the blockchain is recorded as a transaction, and each block contains information, A unique digital signature of the previous block and a time record guarantee the integrity of every block in the chain. Therefore, changing any block's data might potentially jeopardise the integrity of every other block in the chain. Because alterations to metaverse data require the approval of a significant number of blocks in the distributed ledger, using blockchain provides data security

8.6.1.2 Data Storage

Data scientists can work together more quickly because of the decentralized nature of blockchain technology, which expedites the location and classification of data. In the metaverse, blockchain ensures data availability, dependability, and transparency by backing up data in each block. Utilizing a consensus-based distributed ledger strengthens resistance against data copying and manipulation. Nonetheless, further research is needed to tackle delays arising from data replication across the entire chain. While modifying data in a blockchain poses challenges, the potential for a hard fork should be taken into account.

8.6.1.3 Data Sharing

Data sharing benefits a wide range of stakeholders in the metaverse in many ways. People and programmes may collaborate more successfully if they use the same platform, as demonstrated in Ref. [9]. This will enable many different applications to provide users with a better overall experience. Businesses will be able to use the metaverse to undertake data statistics by exchanging information across applications. Data sharing will increase customer knowledge, advertising evaluation, content personalization, content strategy creation, and product development in the metaverse.

Data sharing via centralized exchange platforms carries a considerable risk of exposing data owners' private and sensitive information to the metaverse. When the requirement for real-time data arises in an environment where traditional data sharing is prevalent, data flexibility becomes an issue.

Blockchain technology in the metaverse offers stakeholders access to a decentralized and immutable record of all transactions, particularly in applications related to governance and finance [18]. This heightened data transparency benefits all involved parties. Moreover, data ownership remains entirely in the hands of the owner, and distributed ledger technologies enable efficient data audits, reducing the time and costs associated with data validation. The incorporation of smart contracts enhances data-sharing flexibility as they automate agreement implementations, ensuring instantaneous conclusions without intermediaries or delays. The versatility of blockchain allows smart contracts to be programmed in a wide range of languages, further contributing to their widespread adoption in the metaverse.

8.6.1.4 Data Interoperability

Interoperability will be the primary driver of the metaverse. Diverse applications will have the capability to interact and exchange data within the metaverse. The user is granted a unique set of attributes that can be transferred between virtual worlds with the use of an identity standard. The metaverse's creation will result from the integration of diverse digital worlds. However, existing traditional centralized digital platforms suffer from fragmentation and disorganization. Users face the burden of setting up separate accounts, avatars, hardware, and payment systems to engage in various realms. Additionally, constraints hinder the seamless migration of digital assets like NFTs and avatars to different digital environments. The enclosed nature of virtual spaces further complicates mobility between different areas within the metaverse. These limitations highlight the need for a more unified and interconnected metaverse experience.

Blockchain has the potential to enhance the compatibility of virtual worlds across various metaverses, but additional investigation is required. The primary obstacle to achieving cross-blockchain-enabled metaverse interoperability is the presence of multiple publicly accessible blockchains in different virtual universes that lack standardized communication protocols. Making modifications to address this issue will be challenging, as different platforms offer varying degrees of smart contract capabilities (Figure 8.3).

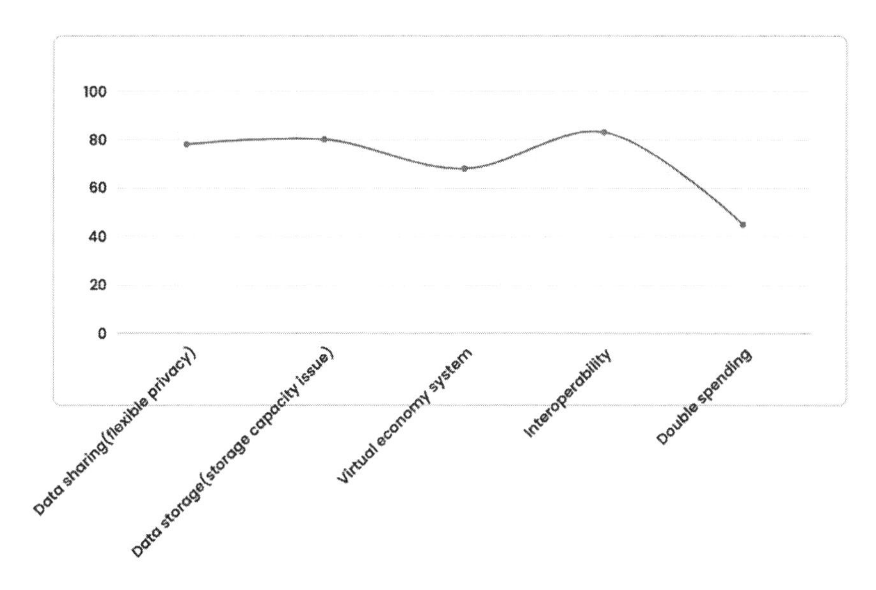

FIGURE 8.3 Improved use cases of blockchain in Metaverse.

8.6.1.5 Data Privacy

The primary obstacle in blockchain technology pertains to human errors and the incorporation of third-party applications that often result in inadequate security measures. Consequently, cyber attackers gain an advantage in compromising user data privacy. Additionally, the public nature of transaction histories exposes asset information and wallet details. While ensuring data transparency in the metaverse was previously unfeasible, this paper proposes a solution by employing various consensus mechanisms such as Proof of Work (PoW), Proof of Authority (PoA), and Proof of Stake (PoS). This approach aims to enhance data security and foster a sense of safety while conducting transactions.

Data that has been recorded cannot be changed or erased without the network's consent, thanks to blockchain's immutability. Enhancing accountability and data integrity within the metaverse, this feature allows for the transparent and auditable tracking of data transactions [12]. Blockchain can enable data minimization, wherein only the information required for certain transactions or interactions is shared throughout the metaverse. By disclosing less unnecessary data, users can protect their privacy to a higher degree.

By making sure your wearable technology is always up to date and virus-free, you can stay one step ahead of hackers. Deep web or black market technologies pose the greatest threat to the security of Metaverse data as of late. It is important to conduct data analysis and establish relevant legal constraints prior to the Metaverse movement becoming widely accepted.

8.6.1.6 Privacy Preserving

Modern human-computer interface (HCI) technology is integrated into the metaverse to allow users to engage socially and fully immerse themselves in virtual settings. Data security becomes increasingly important as Web 3.0, or the Internet of the Metaverse, grows and blurs the lines between the actual and virtual worlds. In order to handle this difficulty, it is important to define the data that has to be collected and to provide openness, particularly when metaverse data might come from a variety of sources.

It is critical to safeguard private data, including biometric information and facial gestures captured by AR/VR devices from users or the metaverse. Gaining control over data is crucial for data management in virtual environments, even if anonymization is used. Incorporating blockchain technology presents viable ways to guarantee a reliable and safe metaverse.

Users can protect their privacy by disclosing less unnecessary information. Data that has been recorded cannot be changed or erased without the network's consent thanks to blockchain's immutability. This functionality can be used to keep track of data transactions in a transparent and auditable manner, improving accountability and data integrity inside the metaverse.

As a result of the huge increase in data generation in the metaverse, apps with weak security safeguards run a higher risk of a data breach.

Users are not anonymous as a result. The creators of the avatar are known to the Metaverse platform owners, so the avatar is not anonymous to them. Avatars, however, can at most be fictitious; moreover, the avatar's in-metaverse activities.

In an internal memo, Meta CTO Andrew Bosworth recognized that "at any meaningful scale, managing people's actions in the metaverse is essentially insolvable," according to The Financial Times. Even if this happened in March 2021, there isn't much reason to be optimistic, especially when one considers Meta's safety record inside its legacy businesses, particularly Instagram. Also, first reports on the security of the metaverse weren't exactly encouraging. According to the Centre for Against Digital Hate, VRChat, a highly regarded programme in Facebook's metaverse, was apparently filled with immorality, prejudice, and sexually explicit material (CCDH).

Blockchain technology empowers metaverse users by granting them data ownership through private and public keys, enabling autonomous data governance. In the blockchain-enabled metaverse, intermediaries are not allowed to exploit or gather data from others, ensuring data owners retain control over their personal information stored in the blockchain. The blockchain's inherent audit trail ensures transaction accuracy and consistency. Additionally, the adoption of zero-knowledge proof allows for easy identification of essential data while preserving individual privacy and ownership rights over metaverse assets. This enhanced user control and security reinforce the attractiveness of blockchain technology for metaverse applications. Moreover, blockchain-based ownership records uniquely connect virtual assets to specific users, significantly reducing the risk of data theft by prohibiting unauthorized access or transfers of digital assets. This section addressed the multiple challenges of collecting, storing, sharing, cooperating, and safeguarding privacy in the metaverse.

8.6.2 Impact of Crisis

Controllers and public authorities are formally beginning to debate whether crimes committed in the metaverse should have an effect in physical world punishment as the metaverse appears to duplicate much of the physical world. The UAE's minister of artificial intelligence, Sultan Al Olama, previously stated that significant metaverse crimes like murder should have an impact on real-world corrections. The simplest system to break all the metaverse security and sequestration enterprises is to enjoin druggies from entering the metaverse. Black and white listing the druggies can do it, as you do in your mobile phone for calls. But, this is the crudest system.

Addressing privacy and security vulnerabilities in the metaverse would necessitate efficient regulation and collaboration among governments, technology corporations, and other stakeholders [10]. However, adopting suitable rules while balancing innovation and user freedoms can be difficult, resulting in arguments and delays in efficiently resolving possible difficulties.

If metaverse users have major privacy and security concerns, participation and engagement may suffer. This, in turn, could have an impact on businesses and creators who rely on the metaverse for revenue. Reduced economic activity in the metaverse may have larger ramifications for companies and individuals who have invested extensively in virtual ventures, and may worsen financial crises [11].

In times of disaster, people may seek sanctuary and distraction in digital settings such as the metaverse. However, if privacy and security difficulties result in undesirable experiences, such as cyberbullying, harassment, or grieving, users' mental health may suffer. The escalation of such episodes may even result in psychological crises for some individuals, who may fail to differentiate between the digital and physical world consequences of such actions.

The metaverse would very certainly become a new battleground for cybercriminals. Malicious actors may utilize security flaws to commit various types of fraud, such as virtual asset theft, bogus item sales, and scams targeting susceptible consumers. These cybercrimes may lead to a sense of instability and disorder inside the metaverse community, hurting overall user participation and trust.

To lessen the impact of privacy and security issues on a crisis, metaverse platforms and developers must prioritize user safety, invest in robust security mechanisms, and create clear standards for user behaviour. Furthermore, cultivating an inclusive and courteous community culture within the metaverse can help to build a safer and more pleasurable virtual environment for all users.

As a result of the New Pails Encrypted White-Listing and Black-Listing Communication On the internet, it is argued that if a product or service is not paid for, then you (or, more particularly, your data) are the product. Social media and networking sites are prime examples of this category.

8.6.3 Resolving Scope

PoS uses a reduced energy than PoW because it doesn't reliant on tricky maths to verify transactions and build new blocks. The metaverse's underlying blockchain network has a less carbon footprint as a result of its energy efficiency.

Secondly, the network's security is also tied to the financial stake that Varyon validators hold as part of a PoS. Validators are economically incentivized to affirm the blockchain's integrity because their stake in the crypto is at risk in the event of disallowed behaviour. This economic rationale improves the security and trustworthiness of the metaverse's blockchain. PoS offers a more decentralized network when compared to PoW. PoW pools a large portion of mining power into big mining operations. In contrast, PoS places this power in the hands of token holders, who can assist in the "validation" and transaction proposal process.

There is a way how the metaverse can address this problem. It is the Proof of Stake consensus mechanism. It can be understood as an alternative to PoW that has distinct several advantages in the virtual world. Unlike the PoW, PoS does not require extensive processing power and does not depend on miners solving the complex computational tasks that validate the transactions and secure the network. It is more affordable and eco-friendly since it consumes much less energy.

In contrast, interaction with PoW has the potential to reduce mining power matching to PoW because mining is mostly handled by a small number of central mining associations. In the event of a PoS metaverse, because almost one corporation might control the blockchain, the top competitor participants are chosen block validators based on the value of their crypto possessions and their willingness to gamble as security.

As they utilize their own cryptocurrency as collateral, validators have "skin in the game". They risk losing the money they staked if they try to validate damaging or fraudulent transactions, which discourages bad behaviour financially.

While PoS does not directly address privacy issues, metaverse platforms have the ability to combine PoS with technologies that improve privacy. Compared to PoW, PoS frequently requires less processing power and specialist hardware. This encourages decentralization and inclusivity by enabling a larger number of individuals to take part in the metaverse network as validators.

PoS networks can often process transactions faster than PoW networks, which makes the metaverse experience more responsive and easy to use [13,14].

Despite PoS's many benefits, no consensus technique is flawless, and there are drawbacks and compromises in every system. Users need robust and secure metaverse environments, which means that privacy features, governance procedures, and security measures need to be well-designed and constantly maintained.

Moreover, different PoS and privacy mechanisms may be implemented in different ways based on the needs of the metaverse platform, its community, and advancements in technology [6]. Research and development in this area will most likely continue to enhance user experience, security, and privacy as the metaverse concept gains traction.

By implementing PoS, the metaverse may create a virtual environment that is more safe, sustainable, and considerate of privacy. This will improve user experience and support the expansion and upkeep of the metaverse ecosystem.

8.7 CONCLUSION

Although there have been significant advancements, the final realization of the lofty concept of a metaverse is still questionable. We are currently seeing encouraging developments, especially in the area of blockchain technology and cryptocurrencies, which support projects like the metaverse and further the integration of blockchain technology into our daily lives. However, achieving a fully-fledged metaverse, an immersive virtual world platform, requires extensive research and development. Notably, the blockchain metaverse is one such ongoing endeavour, leveraging state-of-the-art technology to create a more captivating user experience. While the future of the metaverse is uncertain, progress continues to be made, bringing us closer to a virtual reality that transcends our current digital experiences [20].

There are concerns about cyber security in the metaverse because people are more exposed there than in other parts of the internet. However, that doesn't mean the metaverse shouldn't continue to grow. In fact, it's likely that cyber security concerns will only become more common as the metaverse grows more popular.

Big technology companies are investing in cloud computing and virtual reality companies, because they believe these sectors will grow in the future. BR Softech is providing cryptocurrency development services to help your idea grow more effectively [21,22].

Since the metaverse is a digital world that resembles the real space, the question of how to protect people's personal data is taking on a new level of complexity.

REFERENCES

1. Di Pietro R, Cresci S. Metaverse: Security and privacy issues. In *2021 Third IEEE International Conference on Trust, Privacy and Security in Intelligent Systems and Applications (TPS-ISA)*, 2021, pp. 281–288. IEEE. Atlanta, GA, USA.
2. Huang Y, Li YJ, Cai Z. Security and privacy in Metaverse: A comprehensive survey. *Big Data Mining and Analytics* 2023;6(2):234–247. doi:10.26599/BDMA.2022.9020047.
3. Su Z, Zhang N, Liu D, Luan TH, Shen X. A survey on Metaverse: Fundamentals, security, and privacy.
4. Wang Y, et al. A survey on Metaverse: Fundamentals, security, and privacy. *IEEE Communications Surveys & Tutorials* 2022. 5(1). doi:10.1109/COMST.2022.3202047.
5. Fernandez CB, Hui P. Life, the Metaverse and everything: An overview of privacy, ethics, and governance in Metaverse. In *2022 IEEE 42nd International Conference on Distributed Computing Systems Workshops (ICDCSW)*, Bologna, Italy, 2022, pp. 272–277. doi:10.1109/ICDCSW56584.2022.00058
6. Falchuk B, Loeb S, Neff R. The social Metaverse: Battle for privacy. *IEEE Technology and Society Magazine* 2018;37(2):52–61.
7. Sun J, Gan W, Chao HC, Yu PS. Metaverse: Survey, applications, security, and opportunities. arXiv preprint arXiv:2210.07990, 2022.
8. Gadekallu TR, Huynh-The T, Wang W, Yenduri G, Ranaweera P, Pham QV, da Costa DB, Liyanage M. Blockchain for the Metaverse: A review. arXiv preprint arXiv:2203.09738, 2022.
9. Fu Y, Li C, Yu FR, Luan TH, Zhao P, Liu S. A survey of blockchain and intelligent networking for the Metaverse. *IEEE Internet of Things Journal* 2022;10(4):3587–610.
10. Huynh-The T, Gadekallu T, Wang W, Yenduri G, Ranaweera P, Pham V, Costa DB, Liyanage M. Blockchain for the Metaverse: A review. 2023. 143, pp. 401–419. doi:10.1016/j.future.2023.02.008.
11. Bosworth A, Nick C. Building the Metaverse Responsibly, 2021. https://about.fb.com/news/2021/09/building-the-metaverse-responsibly/.

12. Upadhyay, Utsav & Kumar, Alok & Sharma, Gajanand & Saini, Ashok & Arya, Varsha & Gaurav, Akshat & Chui, Kwok. Mitigating Risks in the Cloud-Based Metaverse Access Control Strategies and Techniques. International Journal of Cloud Applications and Computing. 14. 1-30, 2024. 10.4018/IJCAC.334364.
13. Gaži P, Kiayias A, Russell A. Stake-bleeding attacks on proof-of-stake blockchains. In *2018 Crypto Valley Conference on Blockchain Technology (CVCBT)*, 2018, pp. 85–92. IEEE. Zug, Switzerland.
14. Saad SM, Radzi RZ. Comparative review of the blockchain consensus algorithm between proof of stake (pos) and delegated proof of stake (dpos). *International Journal of Innovative Computing* 2020;10(2).
15. Nguyen CT, Hoang DT, Nguyen DN, Niyato D, Nguyen HT, Dutkiewicz E. Proof-of-stake consensus mechanisms for future blockchain networks: Fundamentals, applications and opportunities. *IEEE Access* 2019;7:85727–85745.
16. Tosh D, Shetty S, Foytik P, Kamhoua C, Njilla L. CloudPoS: A proof-of-stake consensus design for blockchain integrated cloud. In *2018 IEEE 11th International Conference on Cloud Computing (CLOUD)*, 2018, pp. 302–309. IEEE. San Francisco, CA, USA.
17. Chepurnoy A, Duong T, Fan L, Zhou HS. Twinscoin: A cryptocurrency via proof-of-work and proof-of-stake. Cryptology ePrint Archive, 2017.
18. Deuber D, Döttling N, Magri B, Malavolta G, Thyagarajan SA. Minting mechanism for proof of stake blockchains. In *Applied Cryptography and Network Security: 18th International Conference, ACNS 2020*, Rome, Italy, October 19–22, 2020, *Proceedings, Part I* 18 2020, pp. 315–334. Springer International Publishing.
19. Khan SN, Loukil F, Ghedira-Guegan C, Benkhelifa E, Bani-Hani A. Blockchain smart contracts: Applications, challenges, and future trends. *Peer-to-Peer Networking and Applications* 2021;14:2901–2925.
20. Hasan HR, Salah K. Proof of delivery of digital assets using blockchain and smart contracts. *IEEE Access* 2018;6:65439–65448.
21. Poongodi T, Ramya SR, Suresh P, Balusamy B. Application of IoT in green computing. In *Advances in Greener Energy Technologies*, 2020, pp. 295–323. Springer Science and Business Media Deutschland GmbH.
22. Bajpayi P, Sharma S, Gaur MS, AI driven IoT healthcare devices security vulnerability management. In *2024 2nd International Conference on Disruptive Technologies (ICDT)*, Greater Noida, India, 2024, pp. 366–373. doi:10.1109/ICDT61202.2024.10488939.

9 Blockchain-Based Shielding Framework to Enhance Security in Smart Cities

A. Anitha and T. Haritha

9.1 INTRODUCTION

Information and communication technology (ICT) is used in a smart city to enhance operations' efficiency, information communication with the public, the standard of living, and the quality of government services. The US Community Analysis Bureau started using databases, overhead images, and cluster analysis in the 1960s and 1970s to gather information, distribute resources, and generate reports in an effort to enhance services, defend against natural disasters, and reduce inequalities [1]. As a result, the first generation of technology-driven smart cities was created. The initial creation of smart cities was created by technology companies to educate the public about how technology impacts daily life. This led to the creation of the second wave of smart cities (technologically enabled and city-driven), which examined how innovative technologies and other novel ideas could be applied to develop comprehensive solutions. The third phase (citizen-driven) of smart cities eliminates technology providers and city administrators of their authority. Instead, they developed a plan that involved the public and made it simpler for individuals to communicate with the community. Since cities presently hold more than half of the world's population, urbanization requires changes to accommodate the demands of the growing population. However, this has frequently not happened in the current scenario. The excessive number of cities and the burden that they place on their developing infrastructure have caused numerous significant issues.

The most significant network of Internet-connected devices that exchange data is referred to as the Internet of Things, through various software, user interfaces, and information networks, smart cities communicate people to services. This is possible with any device, including automobiles, household electronics, and street cameras. In order to enable both the public and private sectors to become smarter, which will boost the economy and enhance the quality of life, the data acquired from such devices is kept in the cloud or on computers. Edge computing [2] is used by many Internet of Things devices the most significant and helpful data is sent over the Internet. In addition, to incorporate the various advanced technologies such as Application Programming Interfaces (API), Artificial Intelligence (AI), Cloud

 DOI: 10.1201/9781003459835-9

computing solutions, Machine-to-Machine Communication (M2M), and other technologies with IoT systema are implemented to protect, monitor, and prohibit unauthorized entry to the network of the city-wide data infrastructure.

The combination of artificial intelligence, machine learning, and the Internet of Things offers a wide range of benefits. A parking system [3] and intelligent traffic management [4] are two examples of technologies that can help vehicles find parking spaces and enable digital payment. The Hadoop platform tracks the flow of traffic and changes in traffic signals to reduce traffic congestion. Additionally, a digital city's infrastructure may support cab-sharing rides. Utilizing artificial intelligence, smart city features may help with energy conservation [5] and the preservation of the environment; for example, this can be accomplished by dimming streetlights when the roads are empty. With machine learning methods, these innovations in smart grids [6] can enhance operations, maintenance, and planning in response to consumers' fluctuating electrical demands.

In addition to addressing climate change and air pollution [7], advances in smart cities can also be used to manage refuse and keep the city clean through Internet-connected trash collection, recycling containers, and vehicle management systems [8]. Smart communities can also provide safety measures, such as tracking high-risk areas and making use of devices to provide early warnings of flooding, landslides, cyclones, and droughts. Using IoT devices and cloud services, in addition to real-time space management, building health monitoring, and maintenance notifications in smart buildings [9] also permits citizens to report problems, such as security violations, to the government. Sensors can also monitor infrastructure problems, such as broken water pipes [10]. Smart city technology may also increase production, urban agriculture [11], usage of energy, and other industries' productivity. It will be necessary to assure the social, economic, and environmental viability of resources given future population development. Smart cities enable citizens and municipalities to work together on actions and use intelligent technologies to manage the assets and resources of the city.

Smart cities incorporate a network of Internet of Things (IoT) devices and other technological advances to improve their standard of living and boost economic growth. The procedure consists of four steps. Following is the way to collect, analyse, communicate, and implement information to enhance the lives of people in cities. The ICT framework integrates data from connected assets, items, and devices in real time to assist individuals in making more informed decisions. There are numerous advantages to smart cities, though there are also issues that must be resolved. How are smart cities themselves protected? Yes, Cyberattacks, surveillance, and data theft must be defended against in smart cities. Additionally, they must ensure that the information they present is accurate. To maintain the health of smart cities, it is necessary to implement physical data repositories, secure authentication administration, and identification systems [12]. The primary objectives of security can be split down as follows:

1. **Confidentiality**: Data that should not be viewed by unauthorized parties must be kept confidential.
2. **Integrity**: Accurate data must be difficult to manipulate.
3. **Availability**: Accessing real-time information from electronic devices in a secure manner (Figures 9.1 and 9.2).

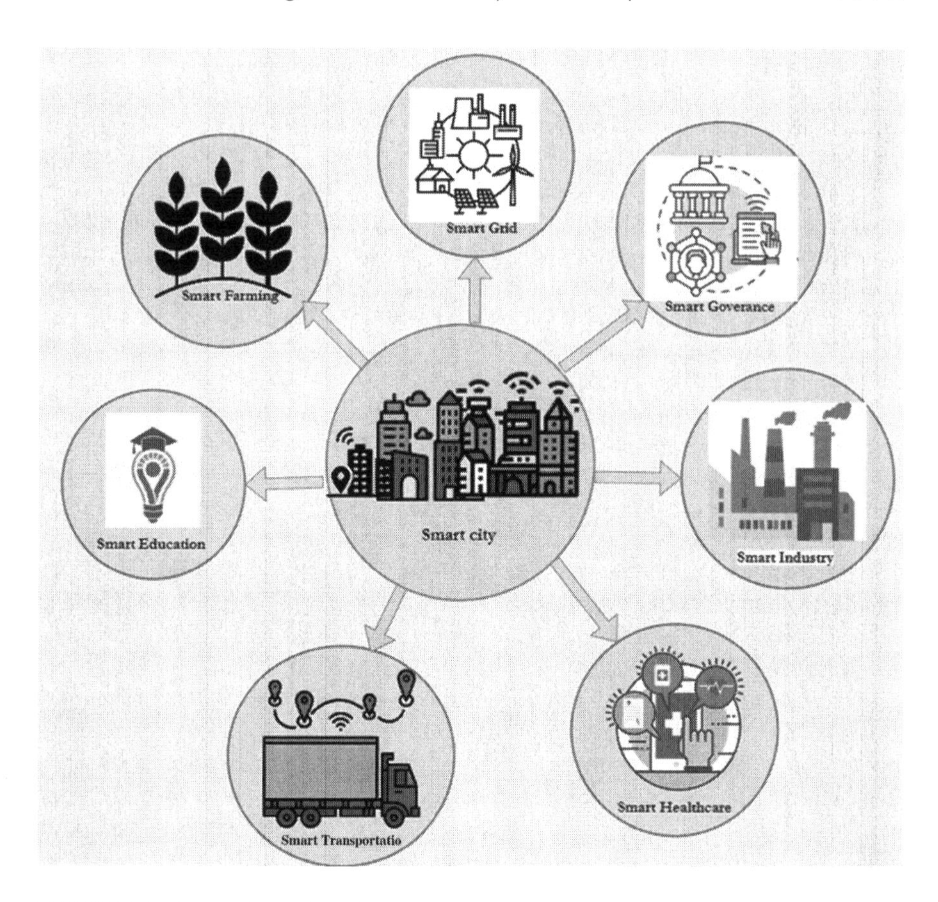

FIGURE 9.1 Smart city applications.

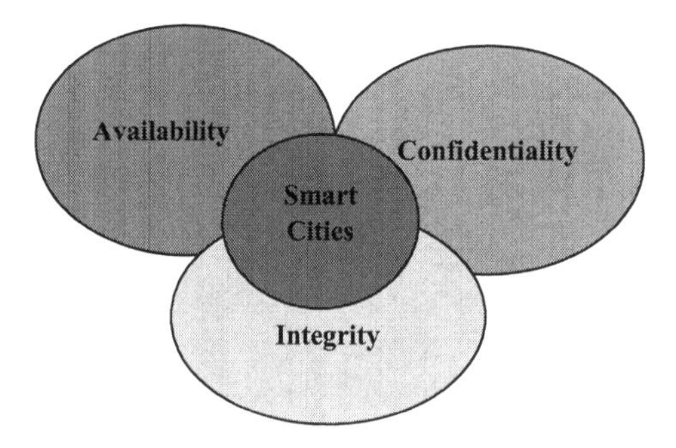

FIGURE 9.2 Triad system of the CIA in smart cities.

Cities have come across confidentiality and security issues in recent years. Smart city services help cities in identifying and implementing new strategies for infrastructure utilization. By providing novel sources of resources and increasing the effectiveness of operations, the alterations assist governments and individuals in enhancing security. In this study, we focus on privacy and security in smart cities from an administrative and technical viewpoint. Determine the numerous security flaws in smart cities and their significance in smart city infrastructure. In addition, we focus on the most recent developments in cybersecurity as well as some well-known attacks that are essential to smart city infrastructure. Finally, we examined the technical solutions against security flaws posed by a variety of attacks to enhance the infrastructure of smart cities.

9.2 RELATED WORKS

This section discusses related work that employs innovative smart methodologies.

9.2.1 Cyber Security Activities in Smart Cities Based on the IoT

Smart cities facilitate the Internet of Things devices for the collection of data and sharing information to central systems like smart meters, streetlights, garbage management, public transportation, and traffic management systems. Cities are now using computing power to control traffic flow, improve how infrastructures work, and create autonomous services for everyday citizens. Digitization has made it easier for cities to collect and manage data by using sensors [13]. The authors in Ref. [14] concentrated on reducing the security challenges at data-centric points in smart cities. Furthermore, the Anomaly Detection-IoT (AD-IoT) method is a smart way to find unusual things. It uses the Random Forest machine learning technique to reduce the hacking data in IoT devices using fog nodes [15]. In smart agriculture, sensors are used to increase the efficiency of data routing and decrease the reduction of transmission signals [16]. Additionally, there is a need to increase the security of the transmission of signals in smart cities. The authors discussed various advancements in cyber security based on IoT Cloud with the limitations of existing deep learning, artificial intelligence, and so on [17]. Moreover, the Internet of Things concentrated on cyberattacks in the field of smart grids. Once IoT devices have been compromised, the power grid becomes vulnerable [18]. In smart homes [19], systems are enabled with IoT devices to improve security issues. To reduce the manual system of monitoring the garbage system, the author [20] implemented IoT sensors in the garbage bin to sense the information through sensors. The incorporation of the Internet of Things (IoT) into various smart technologies in healthcare utilizes sensors to monitor patient conditions at anytime and anywhere to provide better treatments [21]. Moreover, IoT smart sensors are useful for the effective prediction of diseases and easily track the patient's condition with embedded sensor wearables [22]. To improve the smart agriculture systems [23] in urban cities, GPS-based monitoring systems, sensors, intruders, and temperature sensing sensors are used in the farm for the welfare of farmers.

Many of the existing ideas and recommendations are mentioned with the advantages of IoT-enabled devices used to improve end-to-end connectivity, remote monitoring, collection of real-time data, and improve the vision of end customers' experience and operational efficiency in various applications of smart infrastructure. Cities are becoming more secure, quicker, more comfortable, and smarter due to these solutions being used in energy, infrastructure, and automatic services. Smart cities will be the next big thing as IoT becomes more accepted in different areas. There are still major challenges in IoT smart health systems, as the leakage of sensitive patient information through sensor data and failure or power fluctuation of sensor-connected equipment are vulnerable [24]. Even though smart transportation [25] might offer some benefits, it also has a few drawbacks. Cyberattacks are one of the things that people are most worried about. As these vehicles become more connected to the Internet, they could become more vulnerable to hacking, which could frequently disrupt transportation systems or even cause accidents. A data breach happens when unknown people can access user information without the user being aware of it. The person can be manipulated by using this information in the smart manufacturing industry [26].

9.2.2 Cyber Risks in Smart Cities Based on Deep Learning Methods

Smart cities make it slightly simpler for people to live healthily in cities. A smart city seeks to provide quality services through the use of advanced technology and data analytics. Deep Learning (DL) is a form of machine learning that helps users make better decisions by discovering patterns in data. Recent advances in computer vision, such as Edge AI and Deep Learning, a combination of AI and vision with IoT to make the Artificial Internet of Things (AIoT) [27]. These new technologies enable the dealing with large amounts of complex visual data and make it possible for computer vision systems to be fast, reliable, and scalable in real-world applications.

Haque et.al designed a framework to reduce the attacks in sensor measurements of the patient monitoring system using machine learning algorithms [28]. Furthermore, deep-stack neural networks [29] are used to detect malicious activity in an accurate manner that leads to reducing the transmission of data flow and payload in healthcare networks. Smart healthcare systems (SHS) facilitate the ability to monitor and predict patient treatments using machine learning algorithms that are exploited by security vulnerabilities [30]. The advancements in 5G technology with deep learning provide the detection and treatment of diseases in a dynamic way [31]. Additionally, AI-based cyber security techniques such as intrusion detection, anomaly detection, and predictive analysis are used to reduce the risks in smart Industry 4.0 [32]. Cyber-physical systems (CPS) face several challenges, including scalability, centralization, latency, security, and confidentiality. To address these issues, software-defined networking (SDN) and smart infrastructure should be used [33]. Security, private preservation, and robustness are the key challenges in the smart manufacturing industry, and they need to solve the issues with the implementation of federated learning along with optimization techniques [34].

Kalinin et al. assess the various cyber security risks using neural networks associated with sinkhole attacks, grey hole harm, warm hole crimes, Sybil assaults, and illusion attacks in smart infrastructure [35]. The authors focused on improving the cyber threats analysis accuracy using hybrid Deep learning models [36]. To detect cyber threats such as DoS, DDoS, and Ransomware attacks in smart manufacturing industries to overcome the challenges using federated learning-enabled deep learning intrusion detector framework for secure communication [37]. Using decision models in the Industrial Internet of Things (IIOT) to identify malicious nodes and denial of service attacks in the communication, documenting, and storage of data from IoT devices in order to improve security [38]. Organizations have been considering AI-based solutions to identify and eliminate threats as a way to reduce risk.

Artificial intelligence can be beneficial in cyber security because it makes it easier for security experts to analyse, investigate, and understand crimes. It makes the devices that companies use to fight hackers better and helps them securely keep customer information. AI and machine learning are growing increasingly essential for cybersecurity because they can analyse large amounts of data to find threats like fraudulent activities and malware. While handling large amounts of data and analysing the traffic to find any possible threats and identify the unknown threats easily with the nature of the automatic decision-making process. still, cybersecurity has challenges such as involving a lot of data, it's hard for people to maintain fast and accurate in AI and machine learning systems.

Internet of Things (IoT), cloud computing, and combined networks will make it simpler for citizens and local governments to collaborate and generate new ideas, thus improving the smartness of the city. Despite the potential benefits of digitization, it also presents security and privacy issues. We continue to move towards highly digitalized, technology-dependent, and advanced city growth and management. Smart cities are creative places, incorporate technology a lot, and are on the cutting edge of new ideas. The creation and administration of these locations depend highly on recent innovations. This chapter suggests a cyber security system for a smart city that employs smart devices and blockchain technology to establish an encrypted way of communication between individuals.

The rest of the chapter follows as challenges of cyber risks in smart cities are addressed in Section 9.3. Section 9.4 explores the key components based on cyber risks in smart cities, and an analysis of emerging technologies in smart cities is discussed in Section 9.5. The future directions and opportunities for using emerging technologies in smart cities are described in Sections 9.6 and 9.7, which conclude the work.

9.3 CHALLENGES OF CYBER RISKS IN SMART CITIES

Smart cities are comprised of extremely complex networks of interconnected devices, systems, platforms, and individuals. Manufacturers and administrations must secure smart energy, utilities, wastewater, Sanitary items, parking and vehicles, manufacturing and production, construction automation, digital government and telemedicine, monitoring, and security for the public are some of the industries that have benefited from automation [39]. Each day, numerous threats threaten sensible groups. Among these threats are advanced cyberattacks on critical facilities, closing into industrial control systems (ICS), improperly using crime involving low-power wide area networks (LPWAN) and mobile devices, system downtime threats caused by ransomware, controlling data collected by sensors to cause a national crisis (e.g., disaster identifying systems), and the theft of citizen, healthcare, consumer, and personally identifiable information (PII).

The ultimate objective of smart cities is to boost their productivity and efficiency. Nonetheless, if cyber security is not treated seriously, city residents and authorities could be susceptible to major security breaches. Smart communities contain an undetermined number of vulnerabilities and methods for security, and the following are prominent cybersecurity threats:

1. **Man-in-the-Middle Attacks**: These take sensitive information, for instance. The hackers grab a customer's username and password from fake pages and use them on a legitimate one. A harmful emission could result from a "man-in-the-middle" attack on a smart switch in a wastewater system [40].
2. **Malware Attacks**: Also known as "phishing software," destructive software is any programme or file that causes harm to a computer or its user. Different types of malwares include ransomware, Trojan horses, spyware, and viruses [41].
3. **Social Engineering Attacks**: Consists of controlling, affecting, or manipulating a target to gain access to a system or use financial and private data. It employs psychological methods to control users into producing security flaws or revealing confidential data. Fraudulent honey tap and diversion fraud are examples of social engineering attacks [42].
4. Supply chain attacks and threats directed at developers and companies are unique. The objective of hitting app developers is to spread viruses and get the source files, building procedures, and upgrade methods [43]. A cybercriminal could use captured smart meters in a smart city to launch virus attacks on Energy Management Systems (EMS) or steal energy from a city without being detected.
5. **Distributed Denial of Service (DDOS) Attacks**: A cybercrime is when an offender transmits a large amount of Internet data to a server to prevent users from accessing the online services and websites on the server. In smart cities, the most significant risks are disrupted production lines and affected IoT devices' ability to communicate with one another in the smart manufacturing sector [44].
6. **Permanent Denial of Service (PDOS)**: these attacks require the reinstallation or replacement of system hardware. An insecure parking sensor could also be exploited, demanding replacement [45] (Figure 9.3).

9.4 PRINCIPALS OF CYBER SECURITY AND SHIELDING FRAMEWORK COMPONENTS

A cyber security structure, also known as a digital resilience framework, is an array of documents that outline the regulations, operations, and most efficient methods for controlling risks associated with cyber security. It wants to minimize the business's vulnerability to Internet-based crimes by eliminating security flaws.

1. **Risk Analyses**: Assessments of risks related to cyber security assist companies in recognizing, handling, and reducing all forms of cyber risk. Data protection is an essential component of risk prevention techniques and initiatives.

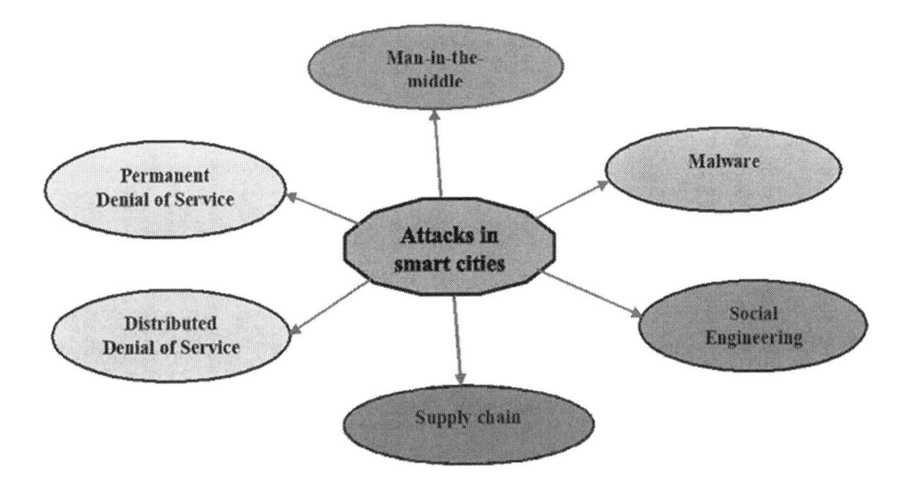

FIGURE 9.3 Multiple attacks in smart cities.

Reduce the identified health risks posed by potentially toxic soil elements in a scenario involving a smart city using of effective management [46].

2. **Security Infrastructure for Network**: Hackers and malicious software that try to take charge of the routing architecture pose the biggest risk to the Internet security of the infrastructure. The intrusion detection systems (IDS), routers, storage systems switches, load balancing devices, domain name systems (DNS), and servers, firewalls encompass the network infrastructure. Through each of these methods, hackers can obtain access to target networks and install malware. Multiple users may utilize the advantages of a network security system for data and protection of networks [47]. In smart cities, the implementation of various threat modelling to create smart digital infrastructure enhances the protection of public data [48] (Figure 9.4).
3. **Endpoint Security**: securing end-user devices against loss of information or leakage, including workstations, laptops, and handheld devices. It contributes to the improvement of security by restricting unidentifiable applications, limiting the accessibility of customers and Internet of Things gadgets, and avoiding the gathering of credentials such as passwords [49]. To mitigate the hazards posed by IoT-enabled end-user devices in an industrial environment in a smart city scenario [50].
4. **Data Security and Prevention**: A secure cyber protection plan saves private, crucial, or confidential data and prevents it from coming into the custody of malicious persons [51], such as company banking details, client account numbers, and employee or user-specific data. Sharing sensitive data in a smart city infrastructure may lead to data breaches or information leakage in the mobile financial services industry [52].

These changes will be made possible by smart city applications and solutions, which will improve the value of lifestyle for city residents. Moreover, smart city upgrades

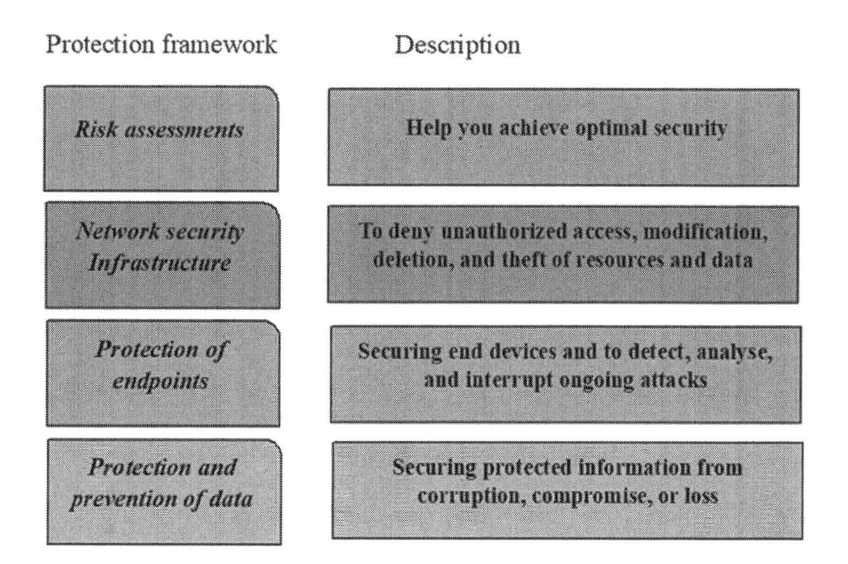

FIGURE 9.4 Cyber security shielding framework.

improve the value of the current facilities. It also contributes to the establishment of fresh revenue sources and efficiency improvements that preserve government agencies and private citizens' funds on a worldwide basis.

9.5 IN-DEPTH ANALYSIS OF NUMEROUS SMART CITY TECHNOLOGIES

9.5.1 Cyber Threats in Cloud Computing-Based Smart Cities

Cloud computing is often safer than working on-site. Most cloud service providers have more security resources than individual enterprises. These cloud service providers maintain up-to-date technology and quickly replace vulnerabilities. However, A number of studies have addressed complex and difficult-to-understand Security issues and cyber threats in cloud computing. The authors discussed in Ref. [53] the structures of cloud computing, as well as security risks, challenges, and approaches to address them, and also explained current implementation methods, cloud services, and cloud design frameworks. The results of this study were used to figure out the field of cloud security. Whereas authors discussed in Ref. [54] emphasized the significance of the security of data, numerous data leaks, and data loss in cloud computing. The author [55] offers a qualitative evaluation of all security issues associated with service architectures like Software as a Service (SaaS), platform as a Service (PaaS), and Infrastructure as a Service (IaaS). Furthermore, the authors [56] identify the security issues that help to reduce the threats to cloud service providers and users in cloud computing. Authors in Ref. [57] implemented to reduce the cyber risks and protected data to access control of data, security measures, and network issues. Ref. [58] explained enhanced security measures against cyberattacks in the cyber-physical system. The authors addressed the issues in the financial industry [59] to reduce the potential cyber security issues in the collaborations of banks and fintech firms. Learning-based Deep-Q-Network (LDQN) was used to protect health data when it was being sent or accessed. This method stops unauthorized entry and investigates malware performance [60].

Incorporating cloud systems into many smart city services makes it easy for people to use city services. Cloud computing platforms use the network to bring together information from many different sources. Cloud computing systems make it possible to use hardware, software, and platforms with the same logic for service. Numerous individuals and organizations have already reduced IT costs and made their processes more efficient by using cloud computing [61]. Even though existing cloud models are better than on-site models in many ways, they can still be vulnerabilities such as data loss, data breaches, and technical failures between internal and external threats while transferring the data and storage information.

9.5.2 Cyber Security Sustainability in Smart Cities with Emerging Blockchain Technology

Nowadays, utilizing sustainable innovations for the network of smart cities to improve the operation of urban systems and encourage sustainable development will enhance the quality of life today. The growing popularity of this technology and its integration into many smart systems have given rise to significant security and privacy concerns, necessitating the development of efficient solutions. Even though the aforementioned innovations have the potential to enhance society as an entire system, [62,63] discovered that almost all smart applications in smart cities have been found to be susceptible to update-based attacks, such as knowledge attacks, collision assaults, surveillance, spam crimes, and sentiment threats. According to Ref. [64], the essential criteria for effective security and privacy include anonymity, accessibility, authorization, confidentiality, and secrecy. Due to their unique characteristics, issues related to privacy and security pose a challenge to the growth of smart city applications.

The following categories of risks have been outlined for smart cities [65]:

i. **Availability Threats**: the (illegal) administration of resources.
ii. **Integrity Threats**: consist of unauthorized data changes,
iii. **Confidentiality Threats**: include inappropriate disclosure of sensitive data;
iv. **Authenticity Threats**: encompasses unauthorized access to sensitive data and resources; and
v. **Accountability Threats**: involve the denial of an entity's message transmission or reception.

9.5.2.1 Blockchain Innovation

Blockchain technology is a decentralized and peer-to-peer ledger platform designed to track transactions, contracts, agreements, and exchanges [66]. Blockchain was developed to facilitate cryptocurrencies, yet it can be utilized in any kind of transaction without a third party. A blockchain advantage is that an adversary must control 51% of

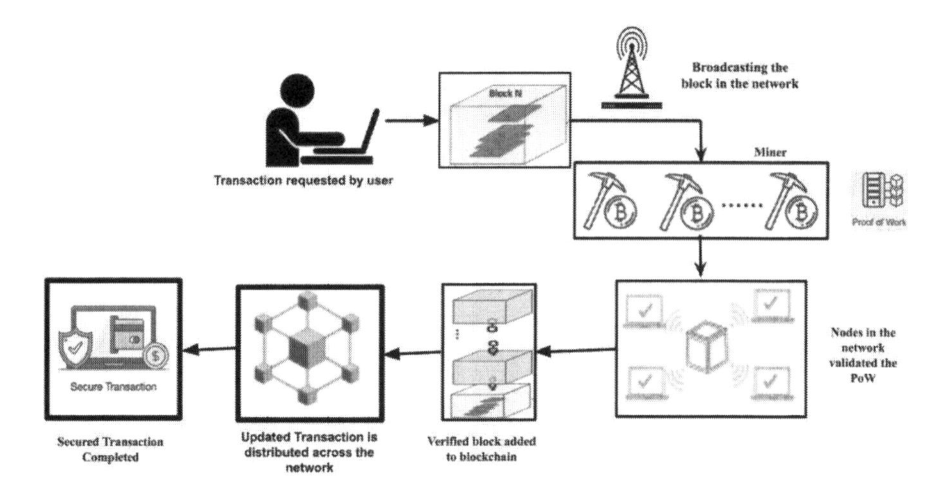

FIGURE 9.5 Transaction validation of user in the network.

the target network's systems to disable its hashing power. As a result, it is technically difficult to attack the blockchain network [67]. The stages involved in the operation of blockchain technology are illustrated in the following scenario (Figure 9.5).

M and D are two components of a smart parking system, and M pays D's space responsibility for parking. The digital representation of this transaction can be represented within a block, which has essential details such as the block number, proof of work, the preceding block, and a comprehensive record of the transaction. The aforementioned portion is distributed to all network nodes. The confirmation and preservation of a transaction in the blockchain occur when a majority of the participating entities validate the block and reach a consensus on its validity. Subsequently, the payment is sent from Company M's account to Authority D's account.

9.5.2.2 How Blockchain Improves Smart Cities Smarter

Since Satoshi Nakamoto's initial concept of cryptocurrencies [68], blockchain technology has advanced significantly. In recent days, blockchain technology has been applicable everywhere, particularly when a trust chain is required. This makes it extremely beneficial for smart cities. Here's an illustration:

i. **Improved Security**: Nowadays, the number of cybercrimes is increasing. according to research, by 2025, companies will be losing approximately USD 10 trillion due to cybercrime. Blockchain could reduce the possibility of leakage. How can smart cities use blockchain technology to enhance security?

   IoT Security: Blockchain can contribute to the security of AI and IoT devices by enabling secure end-to-end encryption, secure communication, and authentication [69].

   Applications Downloads: Blockchain can be employed for verifying the security of upgrades and preventing the execution of harmful applications.

Identity Protection: By employing Blockchain verification methods, it becomes feasible to mitigate personal actions, such as the misplacement of a driver's license or the theft of secret communications, through the utilization of cryptographic techniques [70].

ii. **Preserving the Health care solutions**: Blockchain has numerous benefits in the healthcare industry and is presently undergoing a several transformations. Using Blockchain to establish a distributed network for electronic health records (EHR) [71] and explicit distribution routes for drugs [72] and to avoid the spread of infectious diseases. As blockchain gains popularity, the entire process of diagnosis and therapy [73], including financial transactions, will be able to be conducted on a blockchain.

   Combining Blockchain with advanced technologies, such as AI, can improve the healthcare industry. the integration of AI and Blockchain [74], for example, prevalent medical devices can be utilized to identify a patient's condition. The report is then conveyed using blockchain technology to a physician. The execution of a smart contract can involve the active participation of both the patient and the physician, thereby facilitating the collaborative decision-making process over the appropriate course of action to provide further assistance to the patient, and insurance claims [75] can be submitted using the smart contract. Additionally, telemedicine can assist patients without requiring their physical presence.

iii. **Enhanced Environment Areas**: In smart communities, blockchain technology may help maintain sanitation and hygienic standards. It can provide real-time monitoring of various aspects of waste management. For instance, it provides transparency and immutable information regarding the amount of waste collected, who gathered it, and the way it is recycled or transported. In addition, governments can use blockchain technology to encourage waste management and improve sanitation. This will increase citizen participation and enhance waste management [76].

   Institutions that provide education are required to handle a vast quantity of student data. Likewise, the exchange of information between institutions requires significant time and effort. Blockchain can assist with problem-solving by developing a centralized, immutable record. On the blockchain network, sharing and receiving information between academic institutions is simple. This will help in rendering office management easier [77]. The immutability of blockchain ensures that student data cannot be altered, thereby reducing the potential of forgeries such as faking report cards.

iv. **Additional Funds Savings on Energy**: There are numerous ways in which blockchain can be utilized to assist smart cities in preserving more energy. For example, a network based on blockchain technology might be utilized to track energy consumption. Individuals may additionally provide incentives to other members for their increased support.

9.5.2.3 How Blockchains Are Secure?

Before discussing security issues, let's take a brief look at how it is constructed. The devices contain three extremely vital pieces of data [78]:

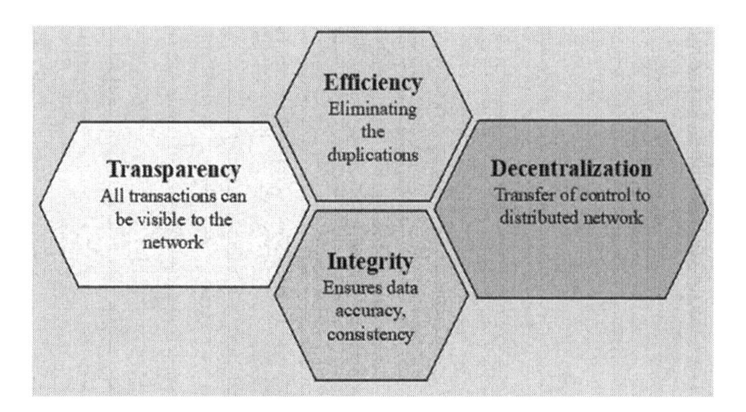

FIGURE 9.6 Features of blockchain.

- Information regarding transactions, such as the time, price, and date.
- A data structure or algorithm-specific code distinguishes between blocks.
- A brief description of all parties to the transaction.

Blockchain technology generates an inherently secure method for managing data. It is founded on cryptography, decentralization, and smart contracts, which ensure the accuracy of all transactions (Figure 9.6).

Secure data platforms, transparent data control, and efficient access procedures, such as biometrics, protect data. Determine how sensors can aid in the protection of vehicles and individuals, as well as in traffic management.

- **Greater Transparency and Interaction**: By applying blockchain-specific features like energy, communications, and security, cities could connect through a single, accessible interconnected system that can exchange data in real time with its residents.
- **Connecting**: Blockchain enables government agencies and individuals to communicate Internet without the need for intermediaries. This would speed up documentation at records, municipal facilities, and other sites.
- **Data Precedes Integrity**: Implementing this technology, you may protect all or a portion of a document so that you are distribute only the relevant portion in a secure, secret, and unmodifiable manner.
- **Efficient Management**: Using blockchain technology, the public and administrators can determine the location and purpose of every resource. Additionally, the former can learn how individuals utilize city services without threatening the privacy of individuals.

9.5.3 Detecting and Preventing Cyber-Attacks Using Blockchain Technology

Globally, 8.4 billion Internet-capable devices appeared in 2017. By 2030, it could reach 500 billion or more [79]. Since all of these devices will be "online," they may be victims of cyberattacks. Cybercriminal activities are becoming increasingly sophisticated, and hackers are employing advanced methods to acquire sensitive data, such as financial information, personal identity information, patient records, and intellectual property.

Cybercriminals also employ DDoS attacks and utilize complex ransomware methods to obtain unauthorized access to information to disrupt business operations. One advantageous component of blockchain technology comes from its ability to bring about transformative changes by providing a decentralized platform for computational purposes. Moreover, it exhibits robustness against current sophisticated attacks. As an outcome of years of cryptography and investigations, blockchain provides an innovative approach for recording information and executing operations, which makes it perfect for environments with rigorous security requirements [80]. Using a central server, the IoT ecosystem of today detects and authenticates individual devices. This offers the ideal opportunity for cybercriminals to employ malicious devices to attack other IoT ecosystem devices [81].

Peer-to-peer (P2P) and trustless Internet of Things (IoT) networks can be developed with blockchain technology, eliminating the need for devices to trust one another and ensuring there is a lack of centralized, single point of risk [82]. Integrating AI and blockchain is also capable of monitoring and predicting cyberattacks, enabling proactive responses to cyber threats. The following procedures may aid in detecting and preventing blockchain-based attacks [9].

- **Eliminating the Human Aspect from Validation**: By removing human intervention from the verification method it removes a potential attack path. By enabling a business's safety system to use a distributed public key infrastructure, blockchain verifies users and devices [83].
- **The Decentralization Nature of Storage**: Each network user is responsible for ensuring that shared and updated data are accurate. This forbids the removal of prevalent data from the blockchain network or the addition of fraudulent data.
- **Traceability**: Every transaction on a blockchain, whether added to a private or public blockchain, is digitally signed and timestamped. This allows organizations to track back every transaction to a specific period and locate the appropriate blockchain entity using the public identifier connected to the transaction.
- **Distributed Denial of Services (DDOS)**: By applying blockchain technology, businesses can decentralize the Domain Name System (DNS), which allows the transmission of data to an enormous number of different nodes and makes it practically impossible to alter the data [84].
- **Data Manipulation**: The network's decentralized architecture, the adoption of cryptography and methods in blockchain, and the close to impossible of anyone to alter the ledger's record all help ensure blockchain consistency, which is essential for preserving an organization's data integrity [85].
- **Improved Handling of Risk Management**: The immutability of blockchain data records improves risk management by maintaining the validity

of authentication and records. In addition, blockchain technology benefits risk management by increasing the transparency of all parties.
- **A Secured IoT**: Using blockchain technology, it is possible to monitor millions of connected devices, and it is also simpler to perform operations and coordinate between devices, resulting in significant cost savings for IoT device manufacturers.

9.6 FUTURE OPPORTUNITIES IN SMART CITIES USING BLOCKCHAIN TECHNOLOGY

Cities can connect with the help of blockchain-specific solutions, such as transportation, energy, or security, through a single open, available, interconnected system that can share data in real time. By creating an immutable, end-to-end encrypted record, blockchain facilitates the prevention of fraud and illegal activities.

Privacy issues can also be resolved on a blockchain by anonymizing personal data and restricting access to it using permissions. Using the Blockchain platform for digital identity solutions empowers individuals and increases the interconnected nature of the entire digital society. One of the primary goals of a Smart City is to improve the quality of life of its citizens through technological innovation.

9.7 CONCLUSION

This survey analyses cyberattacks used in smart cities. We start with a review of previous research and an explanation of the differences between existing technologies and, thus, the motivation for this research. Then, we provide an introduction to blockchain. subsequently, we analysed blockchain-based solutions for the most prevalent cyberattacks in smart city applications, including smart healthcare, smart transportation, smart agriculture, supply chain management, smart infrastructure, and smart residences. Prior to discussing the prerequisites and benefits of blockchain technology for each smart city application, we reviewed the appropriate research and then discussed the necessities for each application. The chapter concludes with an analysis of potential future directions and a mapping of the reviewed solutions to the required specifications. Despite the fact that a number of blockchain-based solutions for smart city applications have been developed, there are still some obstacles to overcome prior to their implementation. The practice of decentralization Storage, efficiency, scalability, and interoperability are among the greatest obstacles that blockchain-enabled solutions must overcome. When presenting a blockchain-based solution, such barriers must be thoroughly considered.

REFERENCES

1. Elmaghraby, A. S., & Losavio, M. M. (2014). Cyber security challenges in Smart Cities: Safety, security, and privacy. *Journal of Advanced Research*, 5(4), 491–497.
2. Anitha, A., & Haritha, T. (2022). The integration of blockchain with IoT in smart appliances: A systematic review. In *Blockchain Technologies for Sustainable Development in Smart Cities* (pp. 223–246). IGI Global.

3. Haritha, T., & Anitha, A. (2023). Asymmetric consortium blockchain and homomorphically polynomial-based PIR for secured smart parking systems. *Computers, Materials & Continua*, *75*(2).
4. Praveen, D. S., & Raj, D. P. (2021). Smart traffic management system in metropolitan cities. *Journal of Ambient Intelligence and Humanized Computing*, *12*, 7529–7541.
5. Ghadami, N., Gheibi, M., Kian, Z., Faramarz, M. G., Naghedi, R., Eftekhari, M., ... & Tian, G. (2021). Implementation of solar energy in smart cities using an integration of artificial neural network, photovoltaic system, and classical Delphi methods. *Sustainable Cities and Society*, *74*, 103149.
6. Tiwari, S., Jain, A., Ahmed, N. M. O. S., Charu, A. L. M., Dafhalla, A. K. Y., & Hamad, S. A. S. (2022). Machine learning-based model for prediction of power consumption in smart grid-smart way towards smart city. *Expert Systems*, *39*(5), e12832.
7. Singh, D. K., Sobti, R., Jain, A., Malik, P. K., & Le, D. N. (2022). LoRa based intelligent soil and weather condition monitoring with internet of things for precision agriculture in smart cities. *IET Communications*, *16*(5), 604–618.
8. Kumar Pulligilla, M., & Vanmathi, C. (2023). An authentication approach in SDN-VANET architecture with Rider-Sea Lion optimized neural network for intrusion detection. *Internet of Things*, *22*, 100723.
9. Bajpayi, P., Sharma, S., & Gaur, M. S. (2024). AI driven IoT healthcare devices security vulnerability management. In *2024 2nd International Conference on Disruptive Technologies (ICDT)*, Greater Noida, India (pp. 366–373). doi:10.1109/ICDT61202.2024.10488939.
10. Zhang, C., Alexander, B. J., Stephens, M. L., Lambert, M. F., & Gong, J. (2023). A convolutional neural network for pipe crack and leak detection in smart water network. *Structural Health Monitoring*, *22*(1), 232–244.
11. Anitha, A., & Acharjya, D. P. (2019). Agriculture crop suitability prediction using rough set on intuitionistic fuzzy approximation space and neural network. *Fuzzy Information and Engineering*, *11*(1), 64–85.
12. Colding, J., Colding, M., & Barthel, S. (2020). The smart city model: A new panacea for urban sustainability or unmanageable complexity? *Environment and Planning B: Urban Analytics and City Science*, *47*(1), 179–187.
13. Ande, R., Adebisi, B., Hammoudeh, M., & Saleem, J. (2020). Internet of Things: Evolution and technologies from a security perspective. *Sustainable Cities and Society*, *54*, 101728.
14. Fan, J., Yang, W., Liu, Z., Kang, J., Niyato, D., Lam, K. Y., & Du, H. (2023). Understanding security in smart city domains from the ANT-centric perspective. *IEEE Internet of Things Journal.*
15. Alrashdi, I., Alqazzaz, A., Aloufi, E., Alharthi, R., Zohdy, M., & Ming, H. (2019). Ad-iot: Anomaly detection of iot cyberattacks in smart city using machine learning. In *2019 IEEE 9th Annual Computing and Communication Workshop and Conference (CCWC)* (pp. 0305–0310). IEEE. Las Vegas, NV, USA.
16. Haseeb, K., Ud Din, I., Almogren, A., & Islam, N. (2020). An energy efficient and secure IoT-based WSN framework: An application to smart agriculture. *Sensors*, *20*(7), 2081.
17. Ahmad, W., Rasool, A., Javed, A. R., Baker, T., & Jalil, Z. (2022). Cyber security in IoT-based cloud computing: A comprehensive survey. *Electronics*, *11*(1), 16.
18. Kimani, K., Oduol, V., & Langat, K. (2019). Cyber security challenges for IoT-based smart grid networks. *International Journal of Critical Infrastructure Protection*, *25*, 36–49.
19. Anitha, A. (2017). Home security system using Internet of Things. *IOP Conference Series: Materials Science and Engineering*, *263*(4), 042026.

20. Anitha, A. (2017). Garbage monitoring system using IoT. *IOP Conference Series: Materials Science and Engineering, 263*(4), 042027.
21. Al-Kahtani, M. S., Khan, F., & Taekeun, W. (2022). Application of Internet of Things and sensors in healthcare. *Sensors, 22*(15), 5738.
22. Khan, M. A. (2021). Challenges facing the application of IoT in medicine and healthcare. *International Journal of Computations, Information and Manufacturing (IJCIM), 1*(1).
23. Suma, N., Samson, S. R., Saranya, S., Shanmugapriya, G., & Subhashri, R. (2017). IOT based smart agriculture monitoring system. *International Journal on Recent and Innovation Trends in computing and communication, 5*(2), 177–181.
24. Khozeimeh, F., Roshanzamir, M., Shoeibi, A., Darbandy, M. T., Alizadehsani, R., Alinejad-Rokny, H., ... & Nahavandi, S. (2022). Importance of wearable health monitoring systems using IoMT; requirements, advantages, disadvantages and challenges. In *2022 IEEE 22nd International Symposium on Computational Intelligence and Informatics and 8th IEEE International Conference on Recent Achievements in Mechatronics, Automation, Computer Science and Robotics (CINTI- MACRo)* (pp. 000245–000250). IEEE.
25. Butler, L., Yigitcanlar, T., & Paz, A. (2020). How can smart mobility innovations alleviate transportation disadvantage? Assembling a conceptual framework through a systematic review. *Applied Sciences, 10*(18), 6306.
26. Ponomareva, L. V., Usacheva, I. V., & Volkova, A. V. (2020). "Smart Manufacturing" in the context of digitalization of business and society. In *Institute of Scientific Communications Conference* (pp. 777–785). Cham: Springer International Publishing.
27. Bhattacharya, S., Somayaji, S. R. K., Gadekallu, T. R., Alazab, M., & Maddikunta, P. K. R. (2022). A review on deep learning for future smart cities. *Internet Technology Letters, 5*(1), e187.
28. Haque, N. I., Rahman, M. A., Shahriar, M. H., Khalil, A. A., & Uluagac, S. (2021). A novel framework for threat analysis of machine learning-based smart healthcare systems. *arXiv preprint arXiv:2103.03472.*
29. Gupta, L., Salman, T., Ghubaish, A., Unal, D., Al-Ali, A. K., & Jain, R. (2022). Cybersecurity of multi-cloud healthcare systems: A hierarchical deep learning approach. *Applied Soft Computing, 118*, 108439.
30. Sundas, A., Badotra, S., Bharany, S., Almogren, A., Tag-ElDin, E. M., & Rehman, A. U. (2022). HealthGuard: An intelligent healthcare system security framework based on machine learning. *Sustainability, 14*(19), 11934.
31. Toding, A., Resha, M., Taliang, A., Rapa, C. I., & Arunglabi, R. (2022). 5G technology in smart healthcare and smart city development integration with deep learning architectures. *International Journal of Communication Networks and Information Security, 14*(3), 99–109.
32. Goyal, S. B., Rajawat, A. S., Solanki, R. K., Zaaba, M. A. M., & Long, Z. A. (2023). Integrating AI with cyber security for smart industry 4.0 application. In *2023 International Conference on Inventive Computation Technologies (ICICT)* (pp. 1223–1232). IEEE.
33. Singh, S. K., Jeong, Y. S., & Park, J. H. (2020). A deep learning-based IoT-oriented infrastructure for secure smart city. *Sustainable Cities and Society, 60*, 102252.
34. Khan, L. U., Alsenwi, M., Yaqoob, I., Imran, M., Han, Z., & Hong, C. S. (2020). Resource optimized federated learning-enabled cognitive internet of things for smart industries. *IEEE Access, 8*, 168854–168864.
35. Kalinin, M., Krundyshev, V., & Zegzhda, P. (2021). Cybersecurity risk assessment in smart city infrastructures. *Machines, 9*(4), 78.

36. Al-Taleb, N., & Saqib, N. A. (2022). Towards a hybrid machine learning model for intelligent cyber threat identification in smart city environments. *Applied Sciences*, *12*(4), 1863.
37. Verma, P., Breslin, J. G., & O'Shea, D. (2022). FLDID: Federated learning enabled deep intrusion detection in smart manufacturing industries. *Sensors*, *22*(22), 8974.
38. Rathee, G., Garg, S., Kaddoum, G., & Choi, B. J. (2020). Decision-making model for securing IoT devices in smart industries. *IEEE Transactions on Industrial Informatics*, *17*(6), 4270–4278.
39. Alibasic, A., Al Junaibi, R., Aung, Z., Woon, W. L., & Omar, M. A. (2017). Cybersecurity for smart cities: A brief review. In *Data Analytics for Renewable Energy Integration: 4th ECML PKDD Workshop, DARE 2016*, Riva del Garda, Italy, September 23, 2016, Revised Selected Chapters 4 (pp. 22–30). Springer International Publishing.
40. Conti, M., Dragoni, N., & Lesyk, V. (2016). A survey of man in the middle attacks. *IEEE Communications Surveys & Tutorials*, *18*(3), 2027–2051.
41. Rudd, E. M., Rozsa, A., Günther, M., & Boult, T. E. (2016). A survey of stealth malware attacks, mitigation measures, and steps toward autonomous open world solutions. *IEEE Communications Surveys & Tutorials*, *19*(2), 1145–1172.
42. Salahdine, F., & Kaabouch, N. (2019). Social engineering attacks: A survey. *Future Internet*, *11*(4), 89.
43. Hudnurkar, M., Deshpande, S., Rathod, U., & Jakhar, S. (2017). Supply chain risk classification schemes: A literature review. *Operations and Supply Chain Management: An International Journal*, *10*(4), 182–199.
44. Saghezchi, F. B., Mantas, G., Violas, M. A., de Oliveira Duarte, A. M., & Rodriguez, J. (2022). Machine learning for DDoS attack detection in industry 4.0 CPPSs. *Electronics*, *11*(4), 602.
45. De Donno, M., Dragoni, N., Giaretta, A., & Mazzara, M. (2018). AntibIoTic: Protecting IoT devices against DDoS attacks. In *Proceedings of 5th International Conference in Software Engineering for Defence Applications: SEDA 2016 5* (pp. 59–72). Springer International Publishing.
46. Yuan, B., Cao, H., Du, P., Ren, J., Chen, J., Zhang, H., … & Luo, H. (2023). Source-oriented probabilistic health risk assessment of soil potentially toxic elements in a typical mining city. *Journal of Hazardous Materials*, *443*, 130222.
47. Verma, A., Prakash, S., Srivastava, V., Kumar, A., & Mukhopadhyay, S. C. (2019). Sensing, controlling, and IoT infrastructure in smart building: A review. *IEEE Sensors Journal*, *19*(20), 9036–9046.
48. Sengan, S., Subramaniyaswamy, V., Nair, S. K., Indragandhi, V., Manikandan, J., & Ravi, L. (2020). Enhancing cyber–physical systems with hybrid smart city cyber security architecture for secure public data-smart network. *Future Generation Computer Systems*, *112*, 724–737.
49. Tan, M. K. S., Goode, S., & Richardson, A. (2021). Understanding negotiated anti-malware interruption effects on user decision quality in endpoint security. *Behaviour & Information Technology*, *40*(9), 903–932.
50. Tedeschi, S., Emmanouilidis, C., Mehnen, J., & Roy, R. (2019). A design approach to IoT endpoint security for production machinery monitoring. *Sensors*, *19*(10), 2355.
51. Li, Q., Wen, Z., Wu, Z., Hu, S., Wang, N., Li, Y., … & He, B. (2021). A survey on federated learning systems: Vision, hype and reality for data privacy and protection. *IEEE Transactions on Knowledge and Data Engineering*.
52. Wewege, L., Lee, J., & Thomsett, M. C. (2020). Disruptions and digital banking trends. *Journal of Applied Finance and Banking*, *10*(6), 15–56.
53. Singh, A., & Chatterjee, K. (2017). Cloud security issues and challenges: A survey. *Journal of Network and Computer Applications*, *79*, 88–115.

54. Kumar, P. R., Raj, P. H., & Jelciana, P. (2018). Exploring data security issues and solutions in cloud computing. *Procedia Computer Science*, *125*, 691–697.
55. Anjana, S., & Singh, A. (2019). Security concerns and countermeasures in cloud computing: A qualitative analysis. *International Journal of Information Technology*, *11*, 683–690.
56. Wani, A. R., Rana, Q. P., & Pandey, N. (2019). Analysis and countermeasures for security and privacy issues in cloud computing. *System Performance and Management Analytics*, 47–54.
57. Mughal, A. A. (2021). Cybersecurity architecture for the cloud: Protecting network in a virtual environment. *International Journal of Intelligent Automation and Computing*, *4*(1), 35–48.
58. Kholidy, H. A. (2021). Autonomous mitigation of cyber risks in the Cyber–Physical Systems. *Future Generation Computer Systems*, *115*, 171–187.
59. Najaf, K., Mostafiz, M. I., & Najaf, R. (2021). Fintech firms and banks sustainability: Why cybersecurity risk matters?. *International Journal of Financial Engineering*, *8*(02), 2150019.
60. Elhoseny, M., Abdelaziz, A., Salama, A. S., Riad, A. M., Muhammad, K., & Sangaiah, A. K. (2018). A hybrid model of Internet of Things and cloud computing to manage big data in health services applications. *Future Generation Computer Systems*, *86*, 1383–1394.
61. Tissir, N., El Kafhali, S., & Aboutabit, N. (2021). Cybersecurity management in cloud computing: Semantic literature review and conceptual framework proposal. *Journal of Reliable Intelligent Environments*, *7*, 69–84.
62. Kitchin, R. (2016). Getting smarter about smart cities: Improving data privacy and data security.
63. Alamer, M., & Almaiah, M. A. (2021). Cybersecurity in Smart City: A systematic mapping study. In *2021 International Conference on Information Technology (ICIT)* (pp. 719–724). IEEE. Amman, Jordan.
64. Mohamed, N., Al-Jaroodi, J., & Jawhar, I. (2020). Opportunities and challenges of data-driven cybersecurity for smart cities. In *2020 IEEE Systems Security Symposium (SSS)* (pp. 1–7). IEEE. Crystal City, VA, USA.
65. Qureshi, K. N., & Iftikhar, A. (2020). Contemplating security challenges and threats for smart cities. In *Security and Organization within IoT and Smart Cities* (pp. 93–118). CRC Press. USA.
66. Zheng, Z., Xie, S., Dai, H. N., Chen, X., & Wang, H. (2018). Blockchain challenges and opportunities: A survey. *International Journal of Web and Grid Services*, *14*(4), 352–375.
67. Nofer, M., Gomber, P., Hinz, O., & Schiereck, D. (2017). Blockchain. *Business & Information Systems Engineering*, *59*, 183–187.
68. Nakamoto, S. (2008). Bitcoin: A peer-to-peer electronic cash system. *Decentralized Business Review*, 21260.
69. Atlam, H. F., & Wills, G. B. (2020). IoT security, privacy, safety and ethics. In *Digital Twin Technologies and Smart Cities* (pp. 123–149). Springer Nature.
70. Toh, C. K. (2020). Security for smart cities. *IET Smart Cities*, *2*(2), 95–104.
71. Yaqoob, I., Salah, K., Jayaraman, R., & Al-Hammadi, Y. (2021). Blockchain for healthcare data management: Opportunities, challenges, and future recommendations. *Neural Computing and Applications*, *34*, 1–16.
72. Humayun, M., Jhanjhi, N. Z., Niazi, M., Amsaad, F., & Masood, I. (2022). Securing drug distribution systems from tampering using blockchain. *Electronics*, *11*(8), 1195.
73. Wang, H. (2020). IoT based clinical sensor data management and transfer using blockchain technology. *Journal of ISMAC*, *2*(03), 154–159.

74. Singh, J., Sajid, M., Gupta, S. K., & Haidri, R. A. (2022). Artificial intelligence and blockchain technologies for smart city. In *Intelligent Green Technologies for Sustainable Smart Cities* (pp. 317–330).
75. Bhawana Kumar, S., Rathore, R. S., Mahmud, M., Kaiwartya, O., & Lloret, J. (2022). BEST— Blockchain-enabled secure and trusted public emergency services for smart cities environment. *Sensors*, *22*(15), 5733.
76. Paturi, M., Puvvada, S., Ponnuru, B. S., Simhadri, M., Egala, B. S., & Pradhan, A. K. (2021). Smart solid waste management system using blockchain and iot for smart cities. In *2021 IEEE International Symposium on Smart Electronic Systems (iSES)(Formerly iNiS)* (pp. 456–459). IEEE. Jaipur, India.
77. Bhushan, B., Khamparia, A., Sagayam, K. M., Sharma, S. K., Ahad, M. A., & Debnath, N. C. (2020). Blockchain for smart cities: A review of architectures, integration trends and future research directions. *Sustainable Cities and Society*, *61*, 102360.
78. Xie, J., Tang, H., Huang, T., Yu, F. R., Xie, R., Liu, J., & Liu, Y. (2019). A survey of blockchain technology applied to smart cities: Research issues and challenges. *IEEE Communications Surveys & Tutorials*, *21*(3), 2794–2830.
79. https://explodingtopics.com/blog/number-of-iot-devices.
80. Ali, O., Jaradat, A., Kulakli, A., & Abuhalimeh, A. (2021). A comparative study: Blockchain technology utilization benefits, challenges and functionalities. *IEEE Access*, *9*, 12730–12749.
81. Miraz, M. H., & Ali, M. (2018). Blockchain enabled enhanced IoT ecosystem security. In *Emerging Technologies in Computing: First International Conference, iCETiC 2018*, London, UK, August 23–24, 2018, Proceedings 1 (pp. 38–46). Springer International Publishing.
82. Sarmah, S. S. (2018). Understanding blockchain technology. *Computer Science and Engineering*, *8*(2), 23–29.
83. Meng, W., Tischhauser, E. W., Wang, Q., Wang, Y., & Han, J. (2018). When intrusion detection meets blockchain technology: A review. *IEEE Access*, *6*, 10179–10188.
84. Saini, K., Kalra, S., & Sood, S. K. (2022). Disaster emergency response framework for smart buildings. *Future Generation Computer Systems*, *131*, 106–120.
85. Das, L., Sharma, S., Yadav, S. A., & Dadhich, K. (2022). Application of blockchain technology in an IoT-integrated framework. In B. Sahoo & S. Yadav (Eds.), *Information Security Practices for the Internet of Things, 5G, and Next-Generation Wireless Networks* (pp. 131–151). IGI Global. doi:10.4018/978-1-6684-3921-0.ch007.

10 Integrating Blockchain, the Internet of Things, and Artificial Intelligence Technologies in Developing Smart Cities

B. Malarvizhi and S. Anusuya

10.1 INTRODUCTION

Smart cities are residential areas that use ICT (information and communication technology) to improve citizen welfare, the calibre of government services, and operational effectiveness. These cities optimize a range of civic services, stimulate economic growth, and improve the general standard of living for citizens by utilizing smart technology and data analysis. Smart city operations involve the use of software, the Internet of Things devices, communication networks, and user interfaces. The IoT devices, such as networked appliances and sensors, collect data, which is then stored on servers or in the cloud. Data processing and analytics, made possible by edge computing and statistical analytics, enable the merging of digital and physical aspects in urban spaces.

Some Characteristics of Smart Cities are:

- Smart Citizen
- Smart Energy
- Smart Technology
- Smart Mobility
- Smart Building
- Smart Healthcare

Smart Citizens: Someone who is tech-savvy and uses it to interact with the Smart City environment, solve neighbourhood problems, and participate in decision-making [4,5].

Smart Energy: Utilizing technology for energy conservation is known as smart energy. It focuses on strong, long-lasting renewable energy sources that lower prices and encourage increased environmental friendliness [13].

 DOI: 10.1201/9781003459835-10

Smart Technology: Remote accessibility or operation from any location, as well as automated or adaptive functionality, are made possible by the capacity to interact and collaborate with other technological networks [1,2].

Smart Mobility: Smart mobility combines all forms of transportation and infrastructure into one cohesive system, utilizing sensors, software, and data platforms to streamline all of these components. These elements include cars—including rideshares, autonomous, and semi-autonomous vehicles—bike shares, traffic signals, buildings, parking spaces, emergency vehicles, and people [14].

Smart Building: Smart buildings, also known as smart facilities, maximize the performance of the structure via the application of technology, services, and systems based on technology for communication and information [61].

Smart Healthcare: A smart healthcare network connects individuals, resources, and healthcare-related institutions while actively managing and intelligently responding to the demands of the medical ecosystem. It does this by utilizing technologies like wearables, the Internet of Things, and mobile Internet to dynamically retrieve data [76,77].

10.1.1 Role of the Internet of Things (IoT) in Developing Smart Cities

The IoT offers limitless possibilities. Big data, artificial intelligence, IoT, and urban data platforms may all work together to make our urban centres smart, sustainable, and effective places with proper planning, deployment, and administration. The collaborative use of information is the key to the success of all industries, including manufacturing, healthcare, and education. Our next generation of smart cities will be more intelligent than ever, thanks to data collection and useful ideas that are put into action [61,62,75].

10.1.2 Role of Blockchain Technology in Developing Smart Cities

Blockchain enhances the integrity of the gathered personal information. It is now simpler for solar-powered houses to automatically swap excess electricity with other grid users by means of smart contracts developed on the blockchain [31–58].

10.1.3 Role of Artificial Intelligence in Developing Smart Cities

AI can assist cities in interacting with the people they serve by examining surveys, social media, and other data to learn about their requirements and preferences. This can assist towns in creating services that better satisfy the demands of their citizens [72,73].

10.1.4 Qualities of a Smart City

A smart city's core is defined by characteristics, themes, and architectural framework. The theme is the cornerstone of the smart city. Here, the infrastructure for running smart cities is provided by the platform. A smart city is composed of certain characteristics. Applications for smart cities rely on four essential elements: intelligence, comfort, quality of life, urbanization, and sustainability. Ecosystems, pollution, energy, and climate change are all connected to sustainability. The goal of citizens' quality-of-life characteristics is to advance their welfare. Technology, infrastructure, and management have all moved from rural to urban settings as a result of urbanization and smarts. "Intelligence" is defined as the desire to improve the financial, environmental, and social standards of cities and their inhabitants. One of the most notable instances of town expansion in recent times is sustainability. The growth of smart cities has been significantly influenced by sustainability.

1. **Smart Economy**: Comprises concerns about a city's economy's ability to compete, such as the city's economic significance in both domestic and foreign markets, entrepreneurship, adaptability in entrepreneurship, innovation, and work and production [1,2].
2. **Smart People**: This is due to a number of factors related to the growth in social investments [3]. This covers the citizens' qualifications, educational attainment, and social diversity. A few writers have contributed to e-learning initiatives, like city-provided online courses, and distance learning has helped advance the field of intelligence [4,5].
3. **Smart Governance**: Smart governance is defined as taking into account all aspects related to civil facilities, administrative transparency, and political participation [6,7].
4. **Intelligent Transportation**: Transportation encompasses the availability of ICTs, accessibility on a local, national, and worldwide scale, as well as cutting-edge, environmentally friendly transportation networks. These elements are crucial to the globalization of today. By emphasizing the shared use of information and communication technology over individual transportation, city planning is the most effective way to provide advanced transportation [8,9].
5. **Smart Space**: Factors pertaining to typical and meteorological conditions, pollution, inventory management, and environmental security are becoming increasingly important. Resource management, renewable energy (breeze, sunlight, etc.), and reducing ecological footprints are all priorities in urban areas. Optimization can also be raised by other innovations [9–11]
6. **Smart Living**: Everything from health to travel to safety to culture, this feature emphasizes enhancing the quality of life [12,13] (Figure 10.1 and Graph 10.1).

10.2 SMART CITY SUPPORTING PLATFORMS AND TECHNOLOGIES AVAILABLE

This is an important subject to discuss when discussing smart cities. Some mutually enabling technologies that were noted throughout the entire literature review are presented in this article.

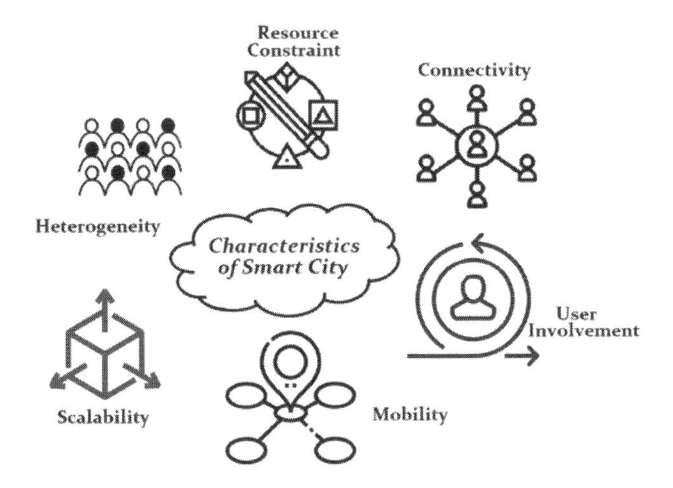

FIGURE 10.1 Characteristics of smart city.

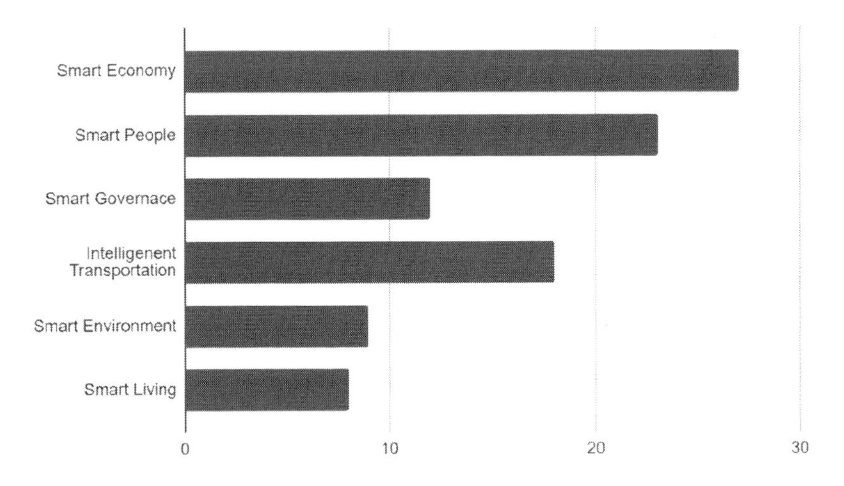

GRAPH 10.1 Smart city scale.

10.2.1 Smart City Technology Enablement

In the process of reviewing the literature, four important technologies for the software platform for smart cities were found. The Internet of Things (IoT), Big Data, Cloud Computing (CC), and Cyber-Physical Systems (CPS) (Figure 10.2).
The technologies supporting smart cities are depicted in Figure 10.3.

Smart cities require the following five technologies: networks, interfaces, processing, detection, and security [59]. The Internet of Things (IoT) and CSPs enable large-scale data organization and exposure, computing distribution, and interface management. Security is essential for all related developments. These four technologies are seen by standards organizations as driving the smart city movement.

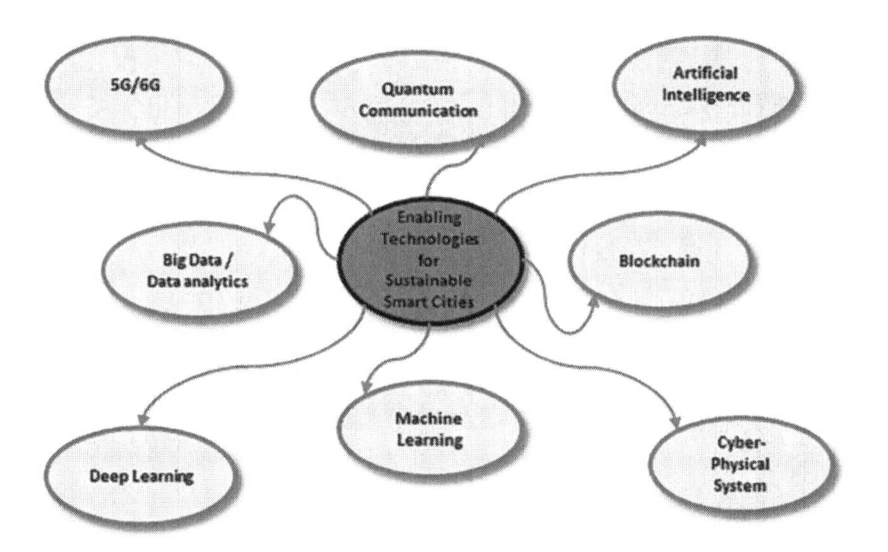

FIGURE 10.2 Sustainable technologies of smart city.

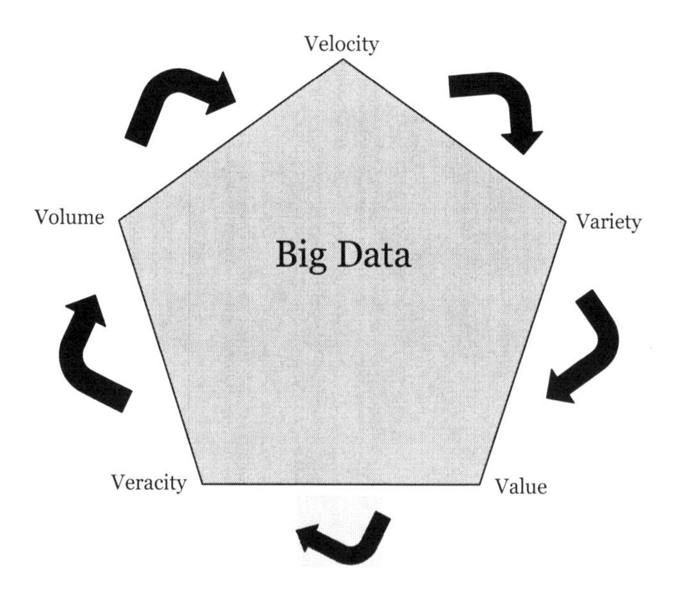

FIGURE 10.3 Characteristics of big data.

10.2.1.1 Internet of Things

The Internet of Things (IoT) explains how items are recognized, assigned unique identifiers, and connected to the Internet in a manner that is network-accessible [60]. The three elements that comprise the Internet of Things environment are sensors, actuators, user-accessible hardware that receives and processes traditional data from the device, and the presentation layer. Hardware includes integrated communication circuits. Middleware for data visualization and manipulation. It is anticipated that smart city platforms will offer similar features. Because a lot of devices are used to collect data about cities. To provide sophisticated smart city administrations, the data collected from these devices must be sent over the network connection for processing and accumulation. In Ref. [16], a few potential uses of IoT in smart cities are discussed. Examples include counting the amount of waste in containers, monitoring traffic lights, assessing noise levels in the city centre, and calculating energy consumption when evaluating the condition of historic buildings. Indicators, a breakdown of smart homes, etc. The Internet of Things and CSP are related technologies. Physical components inside the data framework must be planned, coordinated, checked, and integrated by the CSP [61]. Nonetheless, the infrastructure of the Internet is connected to one or more IoT-connected objects. Despite the fact that these terms can be used interchangeably [62], it is important to distinguish between them due to the notable variations in the platform requirements and objectives.

10.2.1.2 Big and Open Data

Fast and steady should be the "velocity" at which information is handled. It is imperative that directors, administrators, and the city framework possess the ability to promptly address urban issues such as accidents, flooding, and blockages. "Veracity" guarantees because the data is collected and sourced from a range of sources, its quality is important. Unreliable sources can be used to negotiate for the examination of information. Low-quality information sources in cities can include disgruntled sensors, erroneous GPS readings, and malevolent clients. The potential benefits of substantial information to the project are referred to as "value" when considering appropriate variety, the board, and analysis of the information [63]. Large volumes of data generated by urban equipment are now supplied by big data tools in smart cities. The city's temperature, amount of precipitation, and air quality are just a few of the data that are constantly transmitted by the sensor network.

Citizens generate information through cell phones and social networks, and cars like taxis and transports regularly deliver jobs. The concept of open information is another essential component of smart city information [64]. Urban communities can apply creative biological systems to provide fresh solutions to urban problems by bringing together the unrestricted access of citizens, organizations, and non-governmental organizations. Globally, many cities share a variety of datasets related to education, travel, land, and well-being, and that's only the beginning [65]. Applications that benefit citizens, organizations, and regions can be developed by anyone with the necessary creativity and specialized skills.

10.2.1.3 Cyber-Physical Systems

Using register and correspondence techniques, CSPs can be arranged to increase the utility of physical frameworks. As per reference [20], CSP involves balancing computation and physical cycles to enhance physical cycle verification and control. This is achieved by utilizing both nearby and distant PC models on network PCs. CSP is used in a few real-world applications, like control frameworks, smart city communities, and electronic medical devices. The smart city's physical organizational structure is provided by the application Wreck Watch [66], which is intended to detect auto accidents designed for mobile devices can look at the device's GPS and accelerometer to measure the driver's speed both now and in the future. If the data is analysed by the mishap expectation model, Wreck Watch indicates a notable slowdown and notifies the user.

10.2.1.4 Cloud Computing

It provides a very broad, flexible, and easy-to-use framework for processing and archiving information. For intricate smart city frameworks, it is essential. Moreover, it can facilitate the reconfiguration of the framework required for highly potent conditions in intelligent urban areas. The term "Cloud of Things" was coined after several authors [67] examined how cloud computing and IoT were merging. Their plan is to process and store all of the data from the Internet of Things network in a cloud setting, akin to certain other smart city projects [68]. Another idea related to the distributed computing environment of smart cities is a support that supplies sensor data to the cloud computing foundation. In Ref. [69], the authors further developed the idea by referring to it as "Sensing as a Service." Utilizing the idea of programming services, the ClouT platform also defines "City Platform as a Service (CPaaS)" and "City Application Software as a Service (CSaaS)."

10.2.1.5 Technology of Blockchain

The technology that powers cryptocurrencies, known as blockchain, is also a developing platform for developing decentralized applications and data storage. The main idea behind the platform is to provide robust cryptographic guarantees of tamper resistance, immutability, and verifiability while facilitating the process of creating a distributed and duplicated record of data, transactions, and events that are generated by different IT processes [70]. We can most likely ensure these characteristics with the aid of public blockchain systems, even in situations where untrusted users are present in distributed apps with network transaction capabilities. Blockchain technology still has a lot more potential in areas like time management, even though it has gained more notoriety for its use in the implementation of cryptocurrencies like Bitcoin and Ethereum.

10.2.1.6 Digital Twins

A digital twin is an electronic replica or virtual depiction of personnel, systems, devices, locations, processes, and resources. Aircraft engines, cars, and people are among the objects that digital twin technology can mimic. An automotive company generates when it creates a virtual or digital duplicate of a car model, and it is called the digital twin of a real vehicle. If a manufacturer creates a virtual model of its physical production process, that virtual model is a digital twin of the real process. A digital twin is an image that depicts both the current and previous states of a process or physical object. The many facets and dynamics of the Internet of Things device's existence and functioning are shown in this virtual image. Real-time information about the location, state, and/or status of physical assets can be obtained from the digital twin since it is constantly learning and developing. This blending of the physical and digital worlds allows organizations to monitor systems, develop strategies, and anticipate problems before they arise. Digital twins are produced via digital twin technology. This technology integrates network infrastructure graphs, software analytics, artificial intelligence, and the Internet of Things (IoT). The concept of the "smart city" is evident when using digital twins. This technology has the potential to effectively govern cities because of its capacity to plan urban areas and optimize land use. Plan simulation prior to real-time implementation is made possible by digital twins, which helps identify problems early on. Only with a digital twin in place can government agencies effectively evaluate data to improve public life, generate business opportunities, and strengthen ties within the community. Although the idea is still new in many countries, it is expected to gain traction in the next five to ten years [72–74].

10.2.2 Current Platforms in Cities That Are Innovative

It's critical to distinguish between a particular platform selected by a community manager, a general platform for service delivery, and a platform that is used specifically for that purpose when discussing the range of platforms available today. Any city can make use of it [71]. In addition, platforms can be called private or community stages based on whether the engineer can use the stage to increase management. Selling private stages that are available for this kind of situation in different cities is one aspect of the action plan. A number of current platforms must be mentioned on the platform facing the city. These platforms include Webinos, Carriots, Kaa, Sofia2, ICOS, and FIWARE. The following services are provided by these platforms for the installation of city-available infrastructure:

10.2.2.1 Evolution of Smart Cities

Urban areas are the result of dynamic interactions between people and other living forms, encompassing multiple dimensions. Owing to these dynamics and the intricacy of constantly shifting interactions, there is no one definition that fits smart cities. A city's conditions, demands, and circumstances would also vary depending on its socioeconomic culture, living circumstances, intricacies of its people, and interdependencies (Figure 10.4).

10.3 THE SMART CITY CONCEPT

There is no universally accepted definition of what makes a "smart city," so the term has been used in various nations based on social infrastructure, technological advancement, and resident acceptance, among other factors. Other names for these kinds of cities are Future Cities, Green Cities, Digital Cities, Sustainable Cities, etc.

A smart city is defined as one that uses the newest systems and technologies to improve the social lives of its residents. This definition of "smart" cities is straightforward. A smart city is sometimes described as one that has a smart communication network, smart transportation, smart energy, and smart economy are all examples of smart technologies (Table 10.1).

A city that integrates technology more deeply in order to improve the welfare of its residents and protect the environment is considered a "smart city." The city would also develop further with each new technological breakthrough (Figure 10.5).

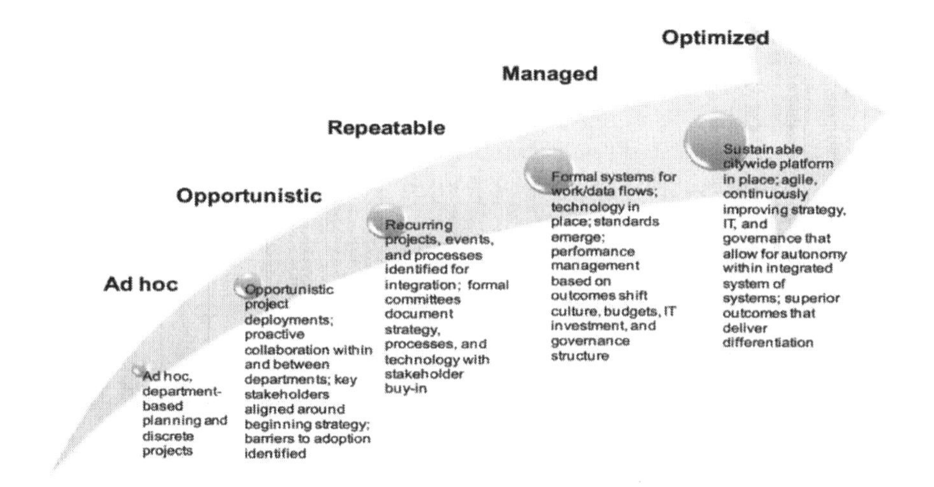

FIGURE 10.4 Evolution of smart cities.

TABLE 10.1
Smart City Concept

Several of the Main Elements of the Current Smart City Concept		
Intelligent learning environment	Bots and Cobots are used extensively	Sophisticated cyber protection
Businesses and virtual companies (wherever possible)	Transport without driver using CAEVs	Elevated degree of human-machine communication
Green energies in smart energy grids	Monitoring of accidents and regulation violations	Precious, fast health care, and assistance
Social, political, and individual freedom	M2M, V2X, R2X, P2X, and IX2X communication and connectivity	Appropriate and successful disaster management system
CAEVs, smart roads, parking signalling, registration, tracking, and other components of the smart mobility network	Maximum recycling level with proper disposal (only when needed)—a sustainable circular economy	Air quality, water, energy, waste, fire safety, and human safety, smart facility management
Super automation and artificial intelligence in IoT	Increasing competency and job opportunities	Appropriate social behaviour for a happy social life to lead a rich and balanced life and seize, The possibilities for the whole social responsibility and smart city development
Enforcing laws at a high level with a virtual and intelligent tax system	Improved communications, transportation, housing, and networking infrastructure	

10.3.1 The Smart Cities of Today

In the 21st century, smart urbanization is becoming a more popular strategy used by cities to innovate. ICT-related smart technologies are often marketed as apolitical, logical, and universal ways to build smart cities without taking into account all of their aspects.

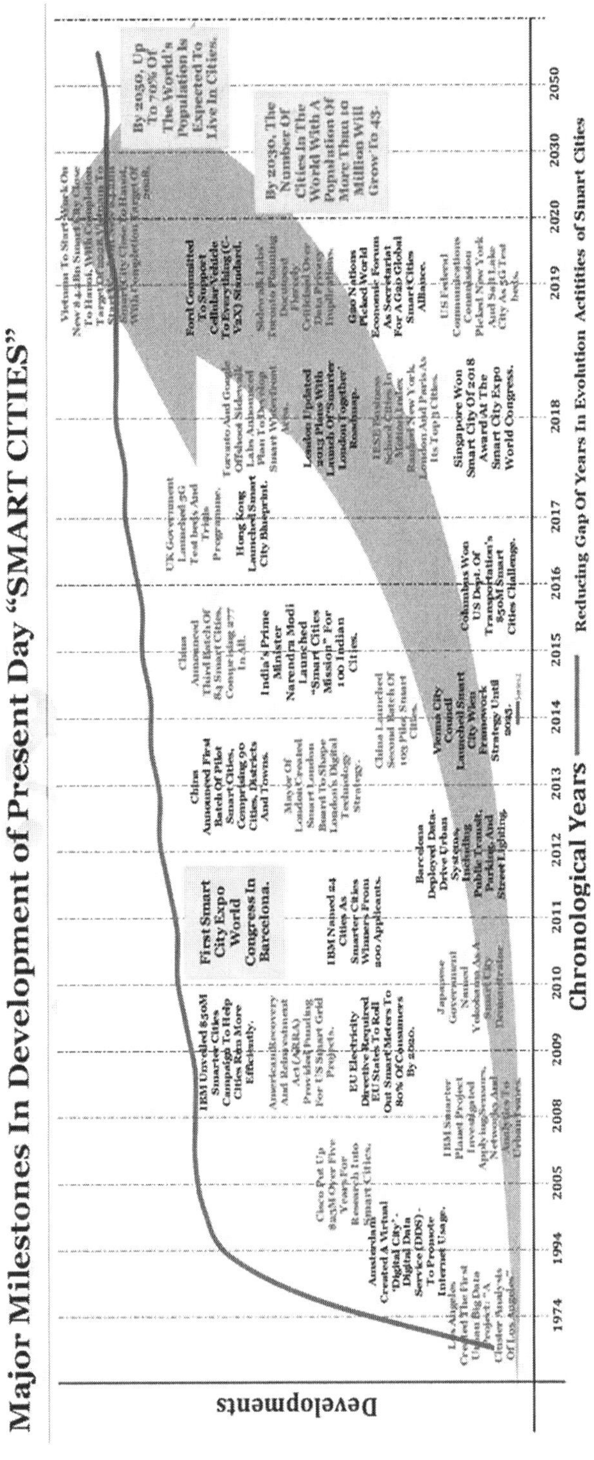

Major Milestones In Development of Present Day "SMART CITIES"

Developments

Los Angeles Created The First Urban Big Data Project: "A Cluster Analysis Of Los Angeles"

Amsterdam Created A Virtual 'Digital City' - Digital Data Service (DDS) - To Promote Internet Usage.

Cisco Put Up $25M Over Five Years For Research Into Smart Cities.

IBM Smarter Planet Project Investigated Applying Sensors, Networks And Analytics To Urban Issues.

IBM Unveiled $50M Smarter Cities Campaign To Help Cities Run More Efficiently.

American Recovery And Reinvestment Act (ARRA) Provided Funding For US Smart Grid Projects.

EU Electricity Directive Required EU States To Roll Out Smart Meters To 80% Of Consumers By 2020.

Japanese Government Named Yokohama As A Smart City Demonstrator.

First Smart City Expo World Congress In Barcelona.

IBM Named 24 Cities As Smarter Cities Winners From 200 Applicants.

Barcelona Deployed Data-Drive Urban Systems, Including Public Transit, Parking, And Street Lighting.

China Announced First Batch Of Pilot Smart Cities, Comprising 90 Cities, Districts And Towns.

Mayor Of London Created Smart London Board To Shape London's Digital Technology Strategy.

China Launched Second Batch Of 103 Pilot Smart Cities.

Vienna City Council Launched Smart City Wien Framework Strategy Until 2025.

China Announced Third Batch Of 84 Smart Cities, Comprising 277 In All.

India's Prime Minister Narendra Modi Launched "Smart Cities Mission" For 100 Indian Cities.

Columbus Won US Dept. Of Transportation's $50M Smart Cities Challenge.

UK Government Launched 5G Test beds And Trials Programme.

Hong Kong Launched Smart City Blueprint.

Toronto And Google Offshoot Sidewalk Labs Announced Plan To Develop Smart Waterfront Area.

London Updated 2013 Plans With Launch Of 'Smarter London Together' Roadmap.

IESE Business School Cities In Motion Index Ranked New York, London And Paris As Its Top 3 Cities.

Singapore Won Smart City Of 2018 Award At The Smart City Expo World Congress.

Ford Committed To Support Cellular Vehicle To Everything (C-V2X) Standard.

Sidewalk Labs' Toronto Planning Document Fiercely Criticised Over Data Privacy Implications.

G20 Nations Picked World Economic Forum As Secretariat For A G20 Global Smart Cities Alliance.

US Federal Communications Commission Picked New York And Salt Lake City As 5G Test beds.

Vietnam To Start Work On New $4.2Bn Smart City Close To Hanoi, With Completion Target Of 2028.

By 2030, The Number Of Cities In The World With A Population Of More Than 10 Million Will Grow To 43.

By 2050, Up To 70% Of The World's Population Is Expected To Live In Cities.

1974 | 1994 | 2005 | 2008 | 2009 | 2010 | 2011 | 2012 | 2013 | 2014 | 2015 | 2016 | 2017 | 2018 | 2019 | 2020 | 2030 | 2050

Chronological Years

Reducing Gap Of Years In Evolution Actitities of Smart Cities

FIGURE 10.5 Present day smart city.

But any ICT application can only deal with problems related to security, surveillance, and resource efficiency; on the other hand, social factors like policymaking, social cohesion, policymaking and citizenship, and behavioural change would require a different kind of long-term, soft smart system. If a society continues to improve the lives of its citizens while utilizing antiquated technologies, it can still be considered intelligent.

Despite all of these complications, a smart city is simply a region that is able to sustain itself, live in harmony with the environment, and efficiently manage its resources (people, buildings, machinery, infrastructure, and natural resources) while employing clean energy sources to make its inhabitants comfortable. A wide definition like this will necessitate the fusion of numerous technical and non-technical concepts, including:

Communication and connectivity for the following scenarios: From everything-to-everything (X2X) to machine-to-machine (M2M) to parking-to-everything (P2X), vehicles-to-everything (V2X) and road-to-everything (R2X).

10.3.2 Smart City Road Map

The four components of the smart city roadmap are as follows: an initial inspection. Figure 10.6 displays the roadmap for smart cities.

Define Community: The links between rural and urban areas and people's movement within them are all included in this, in addition to geography [14,15]

Study of Community: An analysis of the specific development's requirements must be done before choosing to build a smart city. This can be achieved by figuring out how beneficial an initiative like this is. Examine communities

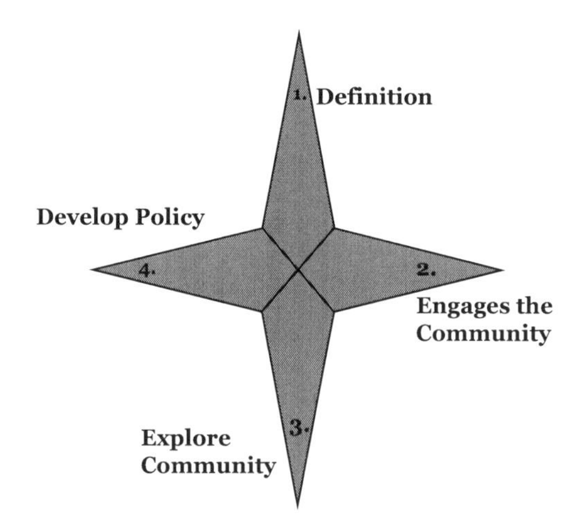

FIGURE 10.6 Road-map of smart city.

to learn about residents, business requirements, and the distinctive characteristics of residents and communities, including age, educational attainment, interests, and urban allure [16–19].

Create Policy for Smart Cities: Establish roles, duties, objectives, and goals. Then, create policies to direct the creation of plans and strategies to reach those goals [20–22].

Interaction between Citizens: Engaging citizens through open data, athletic events, e-government initiatives, and other means can help achieve this [23,24]. The three PPTs—people, process, and technology—are essentially what determine the success of a smart city plan. In order to address their needs, cities must conduct surveys of their residents and communities, comprehend business drivers and procedures, and create policies and objectives that reflect those needs. By using this technology, communities can have their needs met, their quality of life enhanced, and genuine economic opportunities generated [24–26]. This calls for an international, individualized strategy that takes into account local laws, long-term urban planning, and urban culture [27–32].

10.3.3 Smart Cities in Future

Future Smart Cities will emerge from the integration of numerous cutting-edge technologies and solutions from various societal sectors. These cities will be dynamic and ever-changing, posing new challenges and goals as they develop. Cities and societies will continue to be intelligent as long as the new technologies are used to accomplish the goals and address the new challenges (Table 10.2).

Modern technologies can be seamlessly integrated with the city's legacy system in a variety of ways, creating matrixes that better balance their interdependencies. Once created, these matrices will undoubtedly aid in comprehending the complex web of diverse socio-technical elements and the difficulties that come with actually putting things into practice.

TABLE 10.2
Future Day Smart City

Expected Smart Cities: Combining Different Concepts			
Social	Governing	Physical	Economic
Inclusive and integrated	Efficient	Digital	Business-conducive
Protected and safe	Respecting the law	Eco-friendly	Competitive
Knowledgeable and well	Oversaw	Green	Entrepreneurial
Perceptive and engaging	Constructive	Resilient	Creative and effective
Good resources: food, water, air, sanitation, medical	Effectively managed and guided	Sustainable	Productive and resilient

10.3.4 Development and Making of Smart Cities

Only when we determine the technologies and skills required for their architecture, along with their present and future requirements, can we begin to build Smart Cities. DSTA (Draw-See-Think-Arrange) and PDCA (Plan, Do Check, and Action) are two well-known Deming cycles that must be used to maintain focus on the goal during execution. Though the PDCA approach is widely recognized and comprehended, let us attempt to comprehend the DSTA cycle, which is an essential precondition for initiating the planning of a Smart City.

10.4 INDIA'S SMART CITIES

China is expected to lose ground to India as the world's most populous nation soon. India has 16% of the world's population, but only 2.45% of the planet's surface area and 4% of its water resources. It is noteworthy, however, that over the last few decades, there has been a shift toward a rise in the population of cities and a fall in that of rural areas (Figure 10.7 and Graph 10.2).

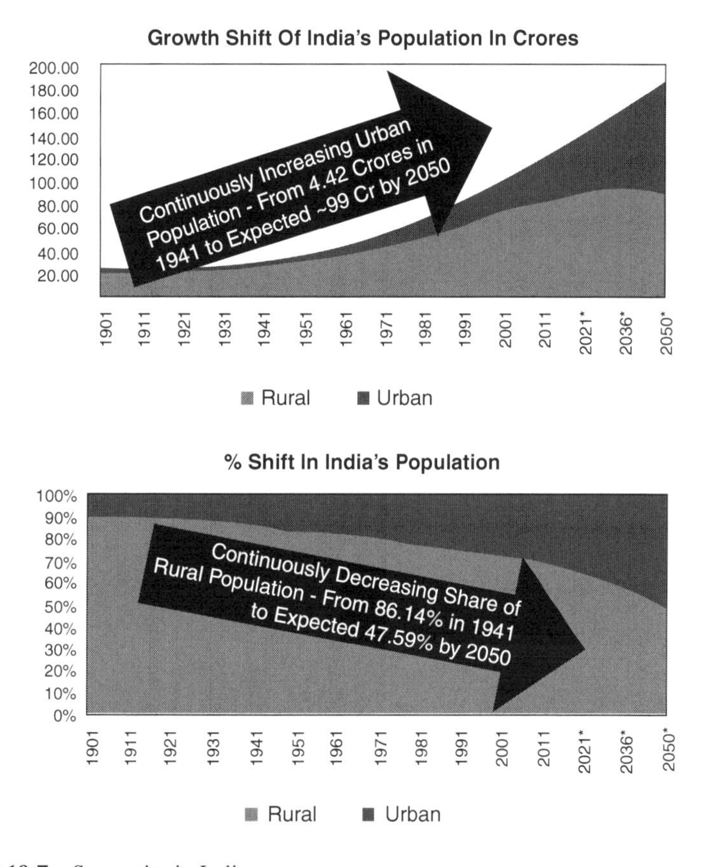

FIGURE 10.7 Smart city in India.

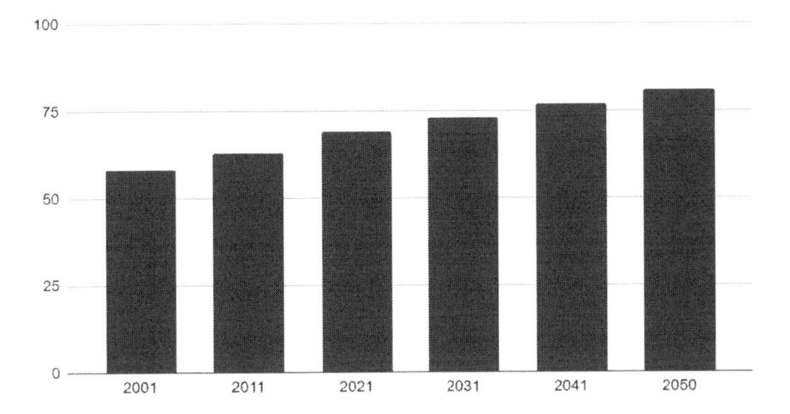

GRAPH 10.2 Growth of India's population.

Since cities are the only places where a country's economy can grow, India's abrupt transition from rural to urban areas will make planned and systematic city growth extremely important.

Currently, 63% of India's GDP is generated by the nearly 31% of the population that lives in urban areas (Census 2011). By 2050, urban areas are predicted to house over 40% of India's population and contribute 75% of the country's GDP due to rising urbanization. Infrastructure development on all fronts—physical, institutional, social, and economic—is necessary for this. A welcome step in that direction is the creation of Smart Cities. However, the idea is very different in India, where it is implemented at varying levels based on the needs of the city in question.

10.4.1 India's Smart City Mission (SCM)

With the intention of improving access for citizens of 110 cities to Prime Minister of India, Shri Narendra Modi, unveiled a massive urban development initiative in 2015 with the goal of providing basic infrastructure and services, a clean environment, and a higher standard of living. The Mission has expanded tremendously over the last six years, completing more than 5,100 projects valued at approximately Rs. 2,05,000 crores that have enhanced the lives of nearly 100 million people. Urban India is currently dealing with a number of problems, including a lack of water, poor air quality, poor sanitation, a disjointed transportation system, rising wealth inequality, unequal distribution, and challenges in accessing public services. India's SCM is focused on generating capital-intensive privatized infrastructure projects and significant pressures from ongoing rural-urban migration. The stress that our cities already experience from the large-scale migration of people from rural areas is being exacerbated by these factors.

10.4.2 Fundamental Ideas of SCM

Citizens at the Core: Every step of the development of a smart city involves citizens.

Greater from Less: The goal of smart cities is to use less energy, money, and other resources while producing greater effects and results.

Competitive and Cooperative Federalism: A system in which states and cities engage in constructive rivalry.

Convergence: Through the convergence of funding sources and initiatives, smart cities aim to create sustainable habitats, an integrated infrastructure and services, and a circular economy.

Technology as a Tool, Not an End in Itself: While technology facilitates and offers speed and scale, it is not the ultimate product of the development of smart cities.

Including Everyone: To be considered "smart," cities must be inclusive of all people, regardless of their age, gender, background, or ability (Figure 10.8).

10.4.3 Smart Solutions under SCM

Presently, the SCM aims to create cities with basic infrastructure, a respectable standard of living for their residents, a clean and sustainable environment, and "smart" solutions to satisfy their needs. The goal is to use compact areas as a model for other ambitious cities, emphasizing inclusive and sustainable development. SCM is a state-of-the-art programme that aims to put significant policies into place to promote the growth of similar cities throughout the nation's different regions and areas.

In order to enhance city planning and, consequently, liveability, the GoI established these minimal definitional boundaries, taking into account the aspirations of its citizens in addition to a desired list of services and infrastructure. The achievable goals for short-, medium-, and long-term goals must be determined, and all cities must collaborate to achieve the same goals.

FIGURE 10.8 Smart solution for Indian smart cities.

10.4.4 Challenges for India

- Vision and leadership combined with decisive action.
- Encourage and support the cooperative and competitive federalism spirit, which is a novel idea in Indian bureaucracy.
- Comprehending the notions of greenfield development, redevelopment, and retrofitting among policymakers, implementers, and other relevant parties.
- Significant time and resource commitment made during the planning stage. This method will not be the same as the traditional DPR-driven method used the government in India.
- Tech-savvy individuals who actively engage in ICT-based governance are necessary for SCM and reforms, especially with regard to mobile tools and the SPV.
- **The Complexity and Scope of Smart City Projects**: The fact that 5,150 projects totalling Rs. 2,05,000 crores will be implemented by the 110 recommended Smart Cities within a five-year period must be acknowledged. Financial innovation is incorporated into the capital investment plans. As of November 2020, 91% of all approved SCM projects had been tendered, and roughly 70% were either finished or in advanced stages of implementation.

10.4.5 Other Concerns

- **Absence of Center-State Coordination**: Appropriate regulation is required for the planning of Smart City development, both vertically and horizontally.
- **Lack of a Master Plan**: The majority of Indian cities lack both a development plan and a master plan. However, a master plan must be created, digitalized, and made available to city planners in order for a Smart City to be developed.
- **Absence of a Timeline**: All plans need to have a completion date and a deadline. The whole Smart City strategy needs to be implemented within a specific period since not all clearances are received on schedule. To make sure that all required approvals for the project are tracked, it is advised that bureaucratic approvals be obtained quickly and online, ideally through a single regulatory body.
- **Insufficient Resources and Expertise**: This is a significant issue in our country, despite the fact that it has not gotten much attention. Large-scale initiatives and programmes require qualified and skilled labour. Considering this, the government is concentrating on the Skill India programme to train individuals with diverse skill sets.
- **Corrupt Practices**: The majority of large-scale projects in India have never been completed or are only partially completed due to corruption (Figure 10.9).

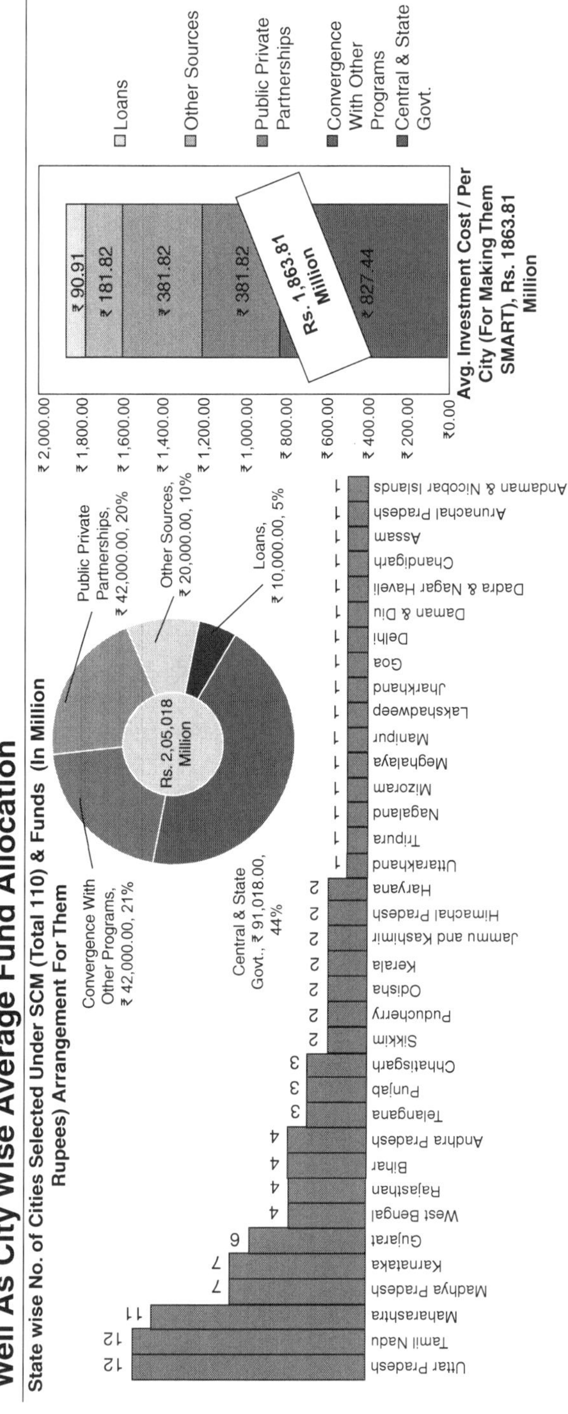

FIGURE 10.9 SCM program and overall city wise average fund.

10.4.6 Planning and Building Sustainable Villages and Cities

Although the government is improving the infrastructure of the 110 cities it has identified to make them "smart," in my opinion, this will never be enough to make them into the perfect "Sustainable Cities" and "Sustainable Villages," not "Smart Cities" as they are known in the West, are what our country actually needs. To make the residents of these cities aware of their SMARTness, we must first concentrate on their local issues, such as employment, corruption, basic educational needs, food, etc. Only then will they be able to improve their lifestyles and create intelligent, self-sustaining, and self-improving cities. Furthermore, every city that is considered "smart" in the modern era must continuously advance to the next level of intelligence. Ultimately, the establishment and maintenance of a Smart City is contingent upon a "smart society"—a society that embraces and uses technology to improve living conditions and vital areas like water, sanitation, transportation, health, and hygiene. Many ancient settlements, such as those in the Indus Valley and Machu Pichu, could be considered Smart Cities of their day due to their sophisticated urban cultures, Outstanding sanitation and water systems, covered street drains, impressive granaries, warehouses, dockyards, brick platforms, and walls that prevented flooding and intruders. While most Indians are still not familiar with the notion of "smart cities," they will eventually learn that smart cities are the product of their people, not their infrastructure. We cannot and should not emulate Western cities, given the sociopolitical and socioeconomic makeup of our country.

10.5 RECENT TRENDS AND CONCEPT

The most recent news from August 2023 is that digital door number plates and QR codes are being pasted on structures in Srinagar, and the smart city's GIS mapping has been completed. The Smart City Project Authorities made the decision to use GIS mapping in order to improve the city's service delivery infrastructure. The programme was a component of the nation's current smart city improvement project. According to the most recent updates, all commercial and residential properties included in the project have been mapped and physically verified. This marked the start of the project as a whole. The authorities are currently pasting the distinct QR codes and Digital Door Numbers linked to these properties as they get ready to begin work on the second phase. The survey for more than 2.25 lakh properties in the city has been finished by the authority. People can use the QR codes that have been pasted on these properties to scan and send the property's address. The Smart City's areas have been classified by the authorities according to various factors such as usage types, descriptions, water sources, electrical connections, and so on. Applications of smart city concepts have been shown to improve lifestyles through the use of intelligent services of superior quality in metropolitan areas with high population densities. The rapid population growth of the world creates a number of social, environmental, and economic issues that have the potential to significantly alter the standard of living and way of life for a large number of people. The idea of a smart city holds the potential to increase awareness and encourage the safest use of green energy in public spaces. These days, popular emerging technologies like blockchain, artificial intelligence, and the internet of things are known for having admirable qualities like automation, decentralization, and security. The Internet of Things makes extensive use of blockchain technology and artificial intelligence. These technologies may be connected through IoT for data collection and presentation, blockchain for operational rule definition and infrastructure, and AI for process and rule optimization.

Information and communication technology (ICT) can therefore play a crucial role in the concept of smart cities.

10.6 CONCLUSION

The study report concludes that residents' standards of living are rising as the idea behind a "smart city" gains traction. Expert studies predict that a large number of people will relocate to cities in the upcoming years. The administrations of the cities will face more difficulties as a result.

A few megacities, like Mumbai in India, are already dealing with the issue. The limited quantity of natural resources on Earth is a topic covered in this paper, as they are depleting daily. Therefore, protecting these resources is turning into a laborious task. Under such circumstances, conserving natural resources through technological means is one option. Smart cities have only one main goal: to make people more productive, to do tasks faster, to spend less money, and to be more sustainable. Various components of smart cities include community, transportation, hospitality, grids, smart healthcare, warehouses, and factories. To accomplish the goals, technologies are used in all of these components. This study offers a methodical overview of the area of smart cities. It starts with a description of the idea's history and development, followed by multiple definitions from different writers. Among other things, a smart city needs to be intelligent, sustainable, urbanized, and high quality of life. We'll talk about the smart city's support pillars and roadmap later. The literature on smart buildings, energy, grids, smart cities, and other related topics has been thoroughly reviewed in this paper. A number of the supporting technologies (Cloud Computing, IoT, Big Data, Machine Learning, etc.), supporting tools (Kaa, Sentilo, IBM Competent Managing Center, CitySDK, Sofia2, etc.), and tools (OMNET++, JSIM Simulator, Cooja Simulator, MATLAB, PySim 5G, etc.) are being used by different researchers throughout the entire literature review. Together, these platforms, technologies, and tools enable the effective implementation of the applications for smart cities. Government officials, companies, and researchers who wish to further the development of smart cities are the target audience for this review paper. Following the review of the literature, this paper addresses the common difficulties developers encounter when putting smart city applications into practice. There are many obstacles in the way, but they also present opportunities in this field. Future directions for Smart City research are being considered after a discussion of some potential research avenues.

About 31% of Indians live in cities, accounting for 63% of the nation's GDP, according to the 2011 Census. Forecasts indicate that by 2050, over 40% of India's population will live in cities, accounting for 75% of the nation's GDP.

REFERENCES

1. V. Stepaniuk, J. Pillai, and B. Bak-Jensen, Battery energy storage management for smart residential buildings, In *Proceedings of the 53rd International Universities Power Engineering Conference (UPEC)*, September 2018, pp. 1–6. Glasgow, UK.
2. J. Huang, F. Qian, A. Gerber, Z. M. Mao, S. Sen, and O. Spatscheck, A close examination of performance and power characteristics of 4G LTE networks, In *Proceedings of the 10th International Conference on Mobile Systems, Applications, and Services (MobiSys)*, UK, 2012, pp. 225–238.
3. X. Ge, S. Tu, G. Mao, and C. X. Wang, 5G ultra-dense cellular networks, *IEEE Trans. Wireless Commun.*, vol. 23, no. 1, pp. 72–79, 2016.
4. M. S. Nazir, F. Alturise, S. Alshmrany, H. M. J. Nazir, M. Bilal, A. N. Abdalla, P. Sanjeevikumar, and Z. M. Ali, Wind generation forecasting methods and proliferation of artificial neural network: A review of five years research trend, *Sustainability*, vol. 12, no. 9, p. 3778, 2020.
5. R. Sanchez-Iborra and M.-D. Cano, State of the art in LP-WAN solutions for industrial IoT services, *Sensors*, vol. 16, no. 5, p. 708, 2016.
6. Y. Khare, V. P. Tiwari, A. B. Patil, and K. Bala, Li-Fi technology, implementations and applications, *Int. Res. J. Eng. Technol.*, vol. 3, no. 4, pp. 1391–1394, 2016.
7. L. Sanchez, L. Muñoz, J. Galache, A. Sotres, J. R. Santana, and V. Gutierrez, SmartSantander: IoT experimentation over a smart city testbed, *Comput. Netw.*, vol. 61, pp. 217–238, 2014.
8. J. Han, J. Pei, and M. Kamber, *Data Mining: Concepts and Techniques*. Amsterdam, The Netherlands: Elsevier, 2011.
9. J. Han, C.-S. Choi, W.-K. Park, I. Lee, and S.-H. Kim, Smart home energy management system including renewable energy based on ZigBee and PLC, *IEEE Trans. Consum. Electron.*, vol. 60, no. 2, pp. 198–202, 2014.
10. B. N. Silva, M. Khan, and K. Han, Internet of Things: A comprehensive review of enabling technologies, architecture, and challenges, *IETE Tech. Rev.*, vol. 35, no. 2, pp. 205–220, 2018.
11. B. Nathali Silva, M. Khan, and K. Han, Big data analytics embedded smart city architecture for performance enhancement through real-time data processing and decision-making, *Wireless Commun. Mobile Comput.*, vol. 2017, pp. 1–12, 2017.
12. B. N. Silva, M. Khan, and K. Han, Integration of big data analytics embedded smart city architecture with RESTful web of things for efficient service provision and energy management, *Future Gener. Comput. Syst.*, vol. 107, pp. 975–987, 2020.
13. U. Zafar, S. Bayhan, and A. Sanfilippo, Home energy management system concepts, configurations, and technologies for the smart grid, *IEEE Access*, vol. 8, pp. 119271–119286, 2020.
14. J. H. Lee, R. Phaal, and S.-H. Lee, An integrated service-device-technology roadmap for smart city development, *Technol. Forecasting Social Change*, vol. 80, no. 2, pp. 286–306, 2013.
15. L. Coetzee and J. Eksteen, The Internet of Things-promise for the future? An introduction, In *Proceedings of the IST-Africa Confernece*, May 2011, pp. 1–9.
16. A. Zanella, N. Bui, A. Castellani, L. Vangelista, and M. Zorzi, Internet of Things for smart cities, *IEEE Internet Things J.*, vol. 1, no. 1, pp. 22–32, 2014.
17. K. Carruthers, Internet of Things and beyond: Cyber-physical systems, *IEEE IoT Newslett.*, vol. 2014, pp. 62–64, 2014.
18. M. Chen, S. Mao, and Y. Liu, Big data: A survey, *Mobile Netw. Appl.*, vol. 19, no. 2, pp. 171–209, 2014.

19. F. Salim and U. Haque, Urban computing in the wild: A survey on large scale participation and citizen engagement with ubiquitous computing, cyber physical systems, and Internet of Things, *Int. J. Hum. Comput. Stud.*, vol. 81, pp. 31–48, 2015.
20. M. Janssen, Y. Charalabidis, and A. Zuiderwijk, Benefits, adoption barriers and myths of open data and open government, *Inf. Syst. Manage.*, vol. 29, no. 4, pp. 258–268, 2012.
21. J. White, S. Clarke, C. Groba, B. Dougherty, C. Thompson, and D. Schmidt, R&D challenges and solutions for mobile cyber-physical applications and supporting internet services, *J. Internet Services Appl.*, vol. 1, no. 1, pp. 45–56, 2010.
22. S. Distefano, G. Merlino, and A. Puliafito, Enabling the cloud of things, In *Proceedings of the Sixth International Conference on Innovative Mobile and Internet Services in Ubiquitous Computing (IMIS-2012)*, July 2012, pp. 858–863.
23. M. Aazam, I. Khan, A. A. Alsaffar, and E.-N. Huh, Cloud of things: Inte- grating Internet of Things and cloud computing and the issues involved, In *Proceedings of the 11th The International Bhurban Conference on Applied Sciences & Technology (IBCAST)*, Islamabad, Pakistan, January 2014, pp. 414–419.
24. P. Eichholtz, N. Kok, and J. M. Quigley, Doing well by doing good? Green office buildings, *Amer. Econ. Rev.*, vol. 100, no. 5, pp. 2492–2509, 2010.
25. O. B. Mora-Sanchez, E. Lopez-Neri, E. J. Cedillo-Elias, E. Aceves-Martinez, and V. M. Larios, Validation of IoT infrastructure for the construction of smart cities solutions on living lab platform, *IEEE Trans. Eng. Manag.*, vol. 68, no. 3, pp. 899–908, 2021. doi:10.1109/TEM.2020.3002250.
26. T. Huang, S. Fu, H. Feng, and J. Kuang, Bearing fault diagnosis based on shallow multi-scale convolutional neural network with attention,' *Energies*, vol. 12, no. 20, p. 3937, 2019.
27. L. Catarinucci, D. De Donno, and L. Mainetti, An IoT-aware architecture for smart healthcare systems, *IEEE Internet Things J.*, vol. 2, no. 6, pp. 515–526, 2015.
28. H. Demirkan, A smart healthcare systems framework, *IT Prof.*, vol. 15, no. 5, pp. 38–45, 2013.
29. V. Stepaniuk, J. R. Pillai, B. Bak-Jensen, and S. Padmanaban, Estimation of energy activity and flexibility range in smart active residential building, *Smart Cities*, vol. 2, no. 4, pp. 471–495, 2019.
30. H. Lund, Definitions, In *Renewable Energy Systems: A Smart Energy Systems Approach to the Choice and Modelling of 100% Renewable Solutions*, H. Lund, Ed. New York, NY: Academic, 384, 2014.
31. B. S. Balaji, P. V. Raja, A. Nayyar, P. Sanjeevikumar, and S. Pandiyan, Enhancement of security and handling the inconspicuousness in IoT using a simple size extensible blockchain, *Energies*, vol. 13, no. 7, p. 1795, 2020.
32. S. Chu and A. Majumdar, Opportunities and challenges for a sustainable energy future, *Nature*, vol. 488, no. 7411, pp. 294–303, 2012.
33. H. Kanchev, D. Lu, F. Colas, V. Lazarov, and B. Francois, Energy management and operational planning of a microgrid with a PV-based active generator for smart grid applications, *IEEE Trans. Ind. Electron.*, vol. 58, no. 10, pp. 4583–4592, 2011.
34. A. R. Boynuegri, B. Yagcitekin, M. Baysal, A. Karakas, and M. Uzunoglu, Energy management algorithm for smart home with renewable energy sources, In *Proceedings of the 4th International Conference on Power Engineering, Energy and Electrical Drives*, May 2013, pp. 1753–1758. Istanbul, Turkey.
35. Y.-S. Son, T. Pulkkinen, K.-D. Moon, and C. Kim, Home energy management system based on power line communication, *IEEE Trans. Consum. Electron.*, vol. 56, no. 3, pp. 1380–1386, 2010.
36. M. Erol-Kantarci and H. T. Mouftah, Wireless sensor networks for cost-efficient residential energy management in the smart grid, *IEEE Trans. Smart Grid*, vol. 2, no. 2, pp. 314–325, 2011.

37. I. Zubizarreta, A. Seravalli, and S. Arrizabalaga, Smart city concept: What it is and what it should be, *J. Urban Planning Develop.*, vol. 142, no. 1, pp. 1–9, 2016.
38. M. P. Efthymiopoulos, Cyber-security in smart cities: The case of Dubai, *J. Innov. Entrepreneurship*, vol. 5, no. 1, pp. 1–16, 2016.
39. G. Li, Y. Wang, J. Luo, and Y. Li, Evaluation on construction level of smart city: An empirical study from twenty Chinese cities, *Sustainability*, vol. 10, no. 9, p. 3348, 2018.
40. R. U. Arora, Financial sector development and smart cities: The Indian case, *Sustain. Cities Soc.*, vol. 42, pp. 52–58, 2018.
41. K. Borsekova, S. Koróny, A. Vaňová, and K. Vitálišová, Functionality between the size and indicators of smart cities: A research challenge with policy implications, *Cities*, vol. 78, pp. 17–26, 2018.
42. R.-M. Soe and O. Mikheeva, Combined model of smart cities and electronic payments, In *The Conference for E-Democracy and Open Government (CeDEM)*, May 2017, pp. 194–205. Krems, Austria, doi:10.1109/CeDEM.2017.11.
43. L. Mora and M. Deakin, *Untangling Smart Cities: From Utopian Dreams to Innovation Systems for a Technology-Enabled Urban Sustainability*. Elsevier, 2019.
44. A. Monzon, Smart cities concept and challenges: Bases for the assessment of smart city projects, In *Proceedings of the IEEE International Conference on Smart Cities and Green ICT Systems (SMARTGREENS)*, May 2015, pp. 1–11. Lisbon, Portugal.
45. D. Lu, Y. Tian, V. Y. Liu, and Y. Zhang, The performance of the smart cities in China—A comparative study by means of self-organizing maps and social networks analysis, *Sustainability*, vol. 7, no. 6, pp. 7604–7621, 2015.
46. A. Kirimtat, O. Krejcar, A. Kertesz, and M. F. Tasgetiren, Future trends and current state of smart city concepts: A survey, *IEEE Access*, vol. 8, pp. 86448–86467, 2020. doi:10.1109/ACCESS.2020.2992441.
47. M. A. Rodriguez-Hernandez, Z. Jiang, A. Gomez-Sacristan, and V. Pla, Intelligent municipal heritage management service in a smart city: Telecommunication traffic characterization and quality of service, *Wireless Commun. Mobile Comput.*, vol. 2019, pp. 1–10, 2019.
48. W. Pan and G. Cheng, QoE assessment of encrypted Youtube adaptive streaming for energy saving in smart cities, *IEEE Access*, vol. 6, pp. 25142–25156, 2018.
49. G. Jia, G. Han, J. Jiang, N. Sun, and K. Wang, Dynamic resource partitioning for heterogeneous multi-core-based cloud computing in smart cities, *IEEE Access*, vol. 4, pp. 108–118, 2016.
50. A. Solanki and T. Singh, Flower species detection system using deep convolutional neural networks, In *Proceedings of the International Conference on Evolving Technologies for Computing, Communication and Smart World (ETCCS*, February 2020, pp. 217–231. Delhi, India.
51. T. Singh, A. Solanki, and A. Nayyar, Multilingual opinion mining movie recommendation system using RNN, In *Proceedings of the International Conference on Computing, Communications, and Cyber-Security (IC4S)*, October 2019, pp. 589–605.
52. J. Medina Quero, M. A. Lopez Medina, A. Salguero Hidalgo, and M. Espinilla, Predicting the urgency demand of COPD patients from environmental sensors within smart cities with high-environmental sensitivity, *IEEE Access*, vol. 6, pp. 25081–25089, 2018.
53. X. Li, Assessment of urban fabric for smart cities, *IEEE Access*, vol. 4, pp. 373–382, 2016.
54. C. M. Park, R. A. Rehman, and B.-S. Kim, Packet flooding mitigation in CCN-based wireless multimedia sensor networks for smart cities, *IEEE Access*, vol. 5, pp. 11054–11062, 2017.
55. C. G. Cassandras, Smart cities as cyber-physical social systems, *Engineering*, vol. 2, no. 2, pp. 156–158, 2016.

56. D. Eckhoff and I. Wagner, Privacy in the smart city: Applications, technologies, challenges, and solutions, *IEEE Commun. Surveys Tuts.*, vol. 20, no. 1, pp. 489–516, 2018.
57. G. Falco, A. Viswanathan, C. Caldera, and H. Shrobe, A master attack methodology for an AI-based automated attack planner for smart cities, *IEEE Access*, vol. 6, pp. 48360–48373, 2018.
58. R. Kulandaivel, M. Kulandaivel, F. Balasubramaniam, L. Al-Turjman, M. Mostarda, M. Ramachandran, and R. Patan, Intelligent data delivery approach for smart cities using road side units, *IEEE Access*, vol. 7, pp. 139462–139474, 2019.
59. R. P. Dameri and C. Benevolo, Governing smart cities: An empirical analysis, *Social Sci. Comput. Rev.*, vol. 34, no. 6, pp. 693–707, 2016.
60. N. Tcholtchev, P. Lammel, R. Scholz, W. Konitzer, and I. Schieferdecker, Enabling the structuring, enhancement and creation of urban ICT through the extension of a standardized smart city reference model, In *Proceedings of the IEEE/ACM International Conference on Utility and Cloud Computing Companion (UCC Companion)*, December 2018, pp. 121–127. Zurich, Switzerland, doi:10.1109/UCC-Companion.2018.00045.
61. H. Li, Y. Liu, Z. Qin, H. Rong, and Q. Liu, A large-scale urban vehicular network framework for IoT in smart cities, *IEEE Access*, vol. 7, pp. 74437–74449, 2019.
62. J. M. Fernández-Güell, S. Guzmán-Araña, M. Collado-Lara, and V. Fernández-Añez, How to incorporate urban complexity, diversity and intelligence into smart cities initiatives, In *Proceedings of the International Smart Cities Conference,* Cham, Switzerland: Springer, 2016, pp. 85–94.
63. A. Komeily and R. S. Srinivasan, Sustainability in smart cities: Balancing social, economic, environmental, and institutional aspects of urban life, In *Smart Cities: Foundations, Principles, and Applications*, 2017, pp. 503–534. Wiley, doi:10.1002/9781119226444.ch18.
64. A. Picon, *Smart Cities: A Spatialised Intelligence.* Wiley, 2015.
65. A. Akande, P. Cabral, P. Gomes, and S. Casteleyn, The Lisbon ranking for smart, sustainable cities in Europe, *Sustain. Cities Soc.*, vol. 44, pp. 475–487, 2019.
66. M. A. Hämäläinen, A framework for a smart city design: Digital transformation in the Helsinki smart city, In *Entrepreneurship and the Community.* Cham, Switzerland: Springer, 2020, pp. 63–86. Mervi Hämäläinen.
67. B. Jedari, F. Xia, H. Chen, S. K. Das, A. Tolba, and AL-M. Zafer, A social-based watchdog system to detect selfish nodes in opportunistic mobile networks, *Future Gener. Comput. Syst.*, vol. 92, no. 3, pp. 777–788, 2019.
68. S. Kolozali, D. Kuemper, R. Tonjes, M. Bermudez-Edo, N. Farajidavar, P. Barnaghi, F. Gao, M. Intizar Ali, A. Mileo, M. Fischer, and T. Iggena, Observing the pulse of a city: A smart city framework for real-time discovery, federation, and aggregation of data streams, *IEEE Internet Things J.*, vol. 6, no. 2, pp. 2651–2668, 2019.
69. B. P. L. Lau, S. H. Marakkalage, Y. Zhou, N. U. Hassan, C. Yuen, M. Zhang, and U.-X. Tan, A survey of data fusion in smart city applications, *Inf. Fusion*, vol. 52, pp. 357–374, 2019.
70. M. Hendy, S. Miniaoui, and H. Fakhry, Towards strategic information & communication technology (ICT) framework for smart cities decision-makers, In *2nd Asia-Pacific World Congress on Computer Science and Engineering (APWC on CSE)*, December 2015, pp. 1–7. doi:10.1109/APWCCSE.2015.7476218.
71. M. Mohammadi, A. Al-Fuqaha, M. Guizani, and J.-S. Oh, Semisupervised deep reinforcement learning in support of IoT and smart city services, *IEEE Internet Things J.*, vol. 5, no. 2, pp. 624–635, 2018.
72. G. Mylonas, A. Kalogeras, G. Kalogeras, C. Anagnostopoulos, C. Alexakos, and L. Munoz, Digital twins from smart manufacturing to smart cities: A survey, *IEEE Access*, vol. 9, pp. 143222–143249, 2021. doi:10.1109/ACCESS.2021.3120843.

73. S. M. E. Sepasgozar, Differentiating digital twin from digital shadow: Elucidating a paradigm shift to expedite a smart, sustainable built environment, *Buildings*, vol. 11, no. 4, p. 151, 2021. doi:10.3390/buildings11040151.
74. F. Dembski, U. Wössner, M. Letzgus, M. Ruddat, and C. Yamu, Urban digital twins for smart cities and citizens: The case study of Herrenberg, Germany, *Sustainability*, vol. 12, no. 6, p. 2307, 2020. doi:10.3390/su12062307.
75. S. A. Yadav, S. Sharma, L. Das, S. Gupta, and S. Vashisht, An effective IoT empowered real-time gas detection system for wireless sensor networks, In *2021 International Conference on Innovative Practices in Technology and Management (ICIPTM)*, Noida, India, 2021, pp. 44–49, doi:10.1109/ICIPTM52218.2021.9388365.
76. P. Bajpayi, S. Sharma, and M. S. Gaur, AI driven IoT healthcare devices security vulnerability management, In *2024 2nd International Conference on Disruptive Technologies (ICDT)*, Greater Noida, India, 2024, pp. 366–373. doi:10.1109/ICDT61202.2024.10488939.
77. V. S. A. Devi, V. H. Raj, B. P. Kavin, E. Gangadevi, B. Balusamy, and S. Gite, An atom quest optimizer for CNN to distinguish IDs in SDN and IoT eco system, In *2024 IEEE International Conference on Computing, Power and Communication Technologies (IC2PCT)*, Greater Noida, India, 2024, pp. 1430–1437. doi:10.1109/IC2PCT60090.2024.10486516.

Index

Note: **Bold** page numbers refer to tables and *italic* page numbers refer to figures.

9781032605968